THE GRUNTS OF WRATH

A MEMOIR EXAMINING MODERN WAR AND MENTAL HEALTH

RONNY BRUCE

TouchPoint Press
Relax. Read. Repeat.

THE GRUNTS OF WRATH:
A Memoir Examining Modern War and Mental Health
By Ronny Bruce

Published by TouchPoint Press
Brookland, AR 72417
www.touchpointpress.com

ISBN: 978-1-956851-44-1

TouchPoint Press books may be purchased in bulk or at special discounts for sales promotions, gifts, fundraising, or educational purposes. For details, contact the Sales and Distribution Staff: info@touchpointpress.com or via fax: 870-200-6702.

Editor: Kaitlyn Kelley
Cover Design: David Ter-Avanesyan, @Ter33Design
Cover Images: © Ronald Bruce

First Edition

Printed in the United States of America.

In Loving Memory

My Mother and Father

Carol and Ronny Bruce

Brothers in Arms

Julian Berisford

Lance Carpenter

Brandon Cox

William Greenleaf

Jeremie Speaker

Toby Stinson

Joshua Weymers

Friend

Lauren Cooper

Thank you ladies: Ms. Williams, Ms. Carlson, Ms. Coghlan, Ms. Kelley; and all the boys from the OG platoon.

Well, here I am, anonymous all right, with guys nobody really cares about. They come from the end of the line, most of 'em. Small towns you never heard of: Pulaski, Tennessee. Brandon, Mississippi. Pork Bend, Utah. Wampum, Pennsylvania. Two years' high school's about it. Maybe if they're lucky, a job waiting for 'em back in a factory. But most of 'em got nothing. They're poor. They're the unwanted. Yet they're fighting for our society and our freedom. It's weird, isn't it? At the bottom of the barrel, and they know it. Maybe that's why they call themselves grunts, cause a grunt can take it, can take anything. They're the best I've ever seen, Grandma. The heart and soul. . . .

—Chris Taylor (Charlie Sheen), *Platoon*

CONTENTS

Prologue i
Chapter 1: Friends, Romans, ATLiens 1
Chapter 2: Back in the Saddle 14
Chapter 3: North to Alaska 22
Chapter 4: Welcome to Malekshay 35
Chapter 5: Every Day is Monday 43
Chapter 6: Man Down, Drive On 59
Chapter 7: Bye-bye Malekshay, F**k You Bergdahl 69
Chapter 8: A New Beginning 76
Chapter 9: Ride 'Til We Die 81
Chapter 10: The Beat Goes On 98
Chapter 11: The Madness of Margah 105
Chapter 12: November Coming Fire 121
Chapter 13: Third Platoon Goes Rolling Along 135
Chapter 14: When Ronny Comes Marching Home 145
Chapter 15: The Finish Line is Margah 162
Chapter 16: The Alabama Freezer 171
Chapter 17: Fight the Powers-that-Be 181
Chapter 18: When Ronny Comes Marching Home Again 192
Chapter 19: The Wolf 204
Chapter 20: The Cost of Bergdahl's Freedom 214
Chapter 21: Redemption 219
Glossary of Acronyms and Terms 231
Characters 234

PROLOGUE

It was 2014, and I was wasting away inside my deepest, darkest personal hell. There was no way out. I was ready to pull the plug.

My chance arrived while driving on a Griffin, Georgia, country road. A vehicle sped up from behind me and proceeded to ride my bumper. When I hit the gas, the vehicle matched me. If I slowed, it slowed. To shake the vehicle, I turned on a desolate two-lane street. But the car turned and continued to follow. Its front bumper kissed my rear. In a fit of rage, I pulled to the side of the road.

Please move along, I thought.

Yet, to my astonishment, the vehicle coasted to the side and parked to my rear. The joke was over.

I exited my vehicle and stalked to the driver's side of the tailgating car. It was a lowered "rice-burner," an old Honda Civic, with a spoiler. Two dudes sat inside. They appeared to be teenage white boys who thought they were gangsters.

Man Pleez.

The driver wore a flat bill hat with aviators perched atop his nose and cheap-looking gold swag around his neck. The passenger was a shaggy-haired kid with a nervous look.

The follower, I mentally named him.

Without breaking stride, I rounded the driver's side door and stomped my right heel into it, placing a dent. Ole' Flat Bill reached into his glove box, pulled a pistol, and pointed it at my chest. The gun fidgeted inside his hands.

"Go on, shoot me!" I shouted.

"C'mon man, let's get the hell outta here," the follower said.

"You don't have the balls, son." I egged him on.

Flat Bill hit reverse and the rice-burner sped off.

"Lil bitch!" I shouted as it sped away.

Walking back to my car, I wept in silence. I wept because I knew I wanted that kid to shoot. Had Flat Bill found his balls and pulled the trigger, I'd be given a pass. Suicide by punk. I'd survived the Taliban, Mujahideen, Al-Qaeda, Chechen mercenaries, and other foreign fighters. But I was ready to meet my demise at the hands of two rednecks on the side of a rural Georgia highway. The local paper would've vilified these hoodlums for murdering a "hero" war veteran. I'd be the victim. My wish to end the torment would be granted.

Days later, I arrived at a crossroad—still eager to die. I laid my deathbed—lounging on my couch with a highball in one hand and a Springfield semiautomatic .40 caliber handgun resting on my lap. The highball was strong – seventy-percent bourbon and thirty-percent Coke. I had one problem though. I didn't have the balls to do it. As I wallowed in my funk, I wished I had the guts that my buddies Clover, Paris, and Stansberry possessed.

Perhaps my subconscious mind would help complete the task?

Yes, my subconscious mind. It had done several things over the years that I wasn't proud of, while my rational mind was blackout drunk. During the 2013 Christmas season, it'd led me on a thirty-mile drive through St. Louis. I found myself on a sidewalk, awakened by the bright bulb of a flashlight, with both mirrors smashed off my vehicle. My subconscious mind had led to being tossed from a bar or two and some check-ins inside a drunk tank. I was asking a lot from it, though. My subconscious mind had led me down silly and fruitless endeavors – "no harm, no foul" type stuff. Yet at that moment, I begged it to finish me.

I guzzled the highballs one after the other with the pistol sitting to my side. My mind combed through everything that'd led me here. I had lived life as a rational thinker who always used logic rather than emotion to guide my decisions. I refused to fall in love in Okinawa, San Diego, and Anchorage because I knew I wouldn't be living in those places forever. I never fathered children with my ex-wife during our short marriage because I didn't feel our i's were dotted and t's were crossed. I'd lived by methodical thinking but, at the moment, logic and rationale were nowhere to be found.

I woke the next morning in a haze. My lap was drenched because the glass in my hand had tilted forward and emptied. The couch reeked of whiskey. My head was smoking but thoughts slowly returned.

I'm alive. Where's my gun?

Despite my dizziness, I stood and searched for the handgun destined to kill me. Upon entering my bedroom, I discovered the case for my Springfield .40, closed and fastened, sitting atop my bed. I opened the case and studied the pistol. The magazine was removed, and the bullets weren't inside. I grabbed the weapon and charged it. No round was ejected since the chamber was emptied. The loaded bullet with Ronny B's name was packed inside its container, invisible amongst the others. I had no recollection of clearing the weapon and placing it inside its case. My subconscious mind had sent a clear message.

Time to live again.

Without knowing my end goal, I organized my thoughts and wrote words on paper. I jotted notes and typed ideas. On and off for years, I dabbled with this side project. Producing a manuscript wasn't my initial goal; writing about my experiences was comforting. It healed. It forced me to tap into the discipline that once kept me motivated and led to moderate success during my earlier years. It gave me purpose.

It was necessary to understand why a guy like me – a former social butterfly, happy-go-lucky, accomplished man who despised sitting at home and thought depression was weak – was mired in a deep funk that culminated with me breathing down the barrel of my own gun. During my twenties, I kicked ass. I received an honorable discharge from the marines, utilized the GI Bill, and earned a bachelor's degree. My name was often on the dean's list at Kennesaw State. I taught world history and current issues in high school for a couple years and was an assistant coach for football and wrestling. Marriage was one of the few mistakes I made, but it was rectified with a call to 1-800-DIVORCE. There's no hate for the ex, we didn't gel, so we split in an amicable way.

Despite all this success, the allure of war was too much. In the summer of 2008, weeks before my 32nd birthday, I knocked on Uncle Sam's door and said *let me back in.* Army infantry this time. It'd been ten years since I last wore the uniform. Back then, it was a peacetime military. The only Combat Action Ribbon I saw was worn by a battalion sergeant major, a crusty old bastard still serving over twenty years after Vietnam. An STD from an Asian whorehouse was my biggest threat in those days. Yet, this second dance with Uncle Sam would be a grind.

As I wrote, I had to explore beyond myself. I had to examine my platoon, company, battalion, and brigade. I had to understand the men. It's been twelve years since the "freedom bird" flew my unit home from the twelve-month deployment to Afghanistan. During these years members of my platoon have caught felonies, sat behind bars, slept under bridges, boozed, drugged, and fought. Some men, too young, are buried underground. Don't get me wrong, there are success stories. There were degrees and advanced degrees earned, high salaries obtained, small businesses started, and high-level government positions filled. Some men remained in the army and were promoted to high ranks or served in elite units. Some redeployed. But the bad seems to outweigh the good.

Allow me to spit some stats. My platoon consisted of thirty men and the company housed a hundred, give or take. A dozen years following the deployment, my hand is full when I count a finger for each suicide by people I knew. I don't have enough fingers and toes to count the number of people who succumbed to depression, alcoholism, drug abuse, homelessness, or crime.

It stings each time I receive a text or call asking – "did you hear about so-and-so?" It's devastating to hear – "so-and-so is dead."

I answered a call from an old platoon buddy one Saturday night and was told, "Dude, had you not answered I was gonna blow my brains out."

I talked him down.

One young man was murdered in Alaska, under shady circumstances.

A text from an old platoon mate last week read, "Don't loan you-know-who $$ if he contacts you cuz he's just gonna buy dope. He's living in his car and has several warrants out for his arrest. His own dad told me he's stolen from him and other members of fam."

A sergeant, our company medic, committed an armed robbery following his discharge. While out on bail, he robbed a bank, which resulted in a twenty-five-year prison sentence. I was shocked. Doc was a quality leader and medic, well-regarded. His life-saving exploits from Operation Iraqi Freedom were legendary. Two deployments, two theatres, twenty-seven months of combat . . . twenty-five years in prison. What went wrong?

Doc's prosecutor said, "This defendant received training to protect and serve our country. He betrayed that public trust and used that training against the citizens he took an oath to defend."

Doc isn't the only Charlie Company guy from the 2009-2010 deployment to spend at least a year in prison.

So, what about me?

Before my re-enlistment, I experienced adult failures for the first time. That brought me to Uncle Sam again – my failures and my curiosity about combat. While most of my peers were knee-deep into careers and playing house in a suburban bubble, ole Ronny B said to hell with that. It wasn't appealing. I needed to feel the fire again.

Despite earning a degree and being eligible to serve as an officer, I chose to reenlist as an enlisted grunt. Infantry enlisted men are the trailer park in town compared to the Chateau Elan Winery and Resort of the officer corps.

Events are told from a grunt's-eye-view. No reporters documented our moves. There were no epic battles and no decorated soldiers to fawn over. However, there are several Purple Hearts since half my platoon was wounded in battle. If we'd counted concussions as medals, then Purple Hearts would blanket the platoon. There was nothing elite about us either. None of us were special operators or high-speed war machines. We were grunts on the frontlines.

I consulted with my former acting platoon sergeant, squad leader, team leader, and a few guys from the platoon to ensure authenticity. Events derive from my notes, emails, records, awards, and memory. I added accounts from other soldiers since I'm human and not omnipresent. Yes, I received the blessings of the men. They encouraged me to see this through to the end since the story informs the world community of what went down in the fight against insurgents, guerillas, jihadists, and hardened Taliban armies – along with the price paid. If any mistakes were made, I apologize. The chaos and confusion of war can lead to fuzzy details in some cases.

The characters are real. I used nicknames and first names to protect their identities, or I altered first or last names when referring to them.

Politics are a buzzkill. But discussing politics can't be avoided when telling stories of combat. Politicians and commanders create foreign policy and give orders while soldiers follow orders on behalf of the US Government. Any opinions are my own and don't necessarily reflect those of others in the platoon.

Struggle is a curse most people wouldn't wish for, but it's been a blessing to me. Enduring chaos and accomplishing difficult tasks have value. Failures either break you or lead you to learn and improve. Too many good people aren't dealing with their struggles and failures well. They're lost and hopeless. So, I offer my story, along with the stories of other warriors, to examine struggle and failure—to ask why we struggle.

Perhaps answers and coping mechanisms will be derived and redemption achieved. I speak candidly about my issues to lift others who may feel like a square peg failing to fit into society's round hole. I'm a square peg. The men of my former platoon are square pegs. We are the grunts of wrath.

CHAPTER 1: FRIENDS, ROMANS, ATLIENS

"Cheers," the bartender poured another drink.

I sipped a highball as fresh ink dried on my divorce papers. It wasn't a celebratory night; I wanted to decompress and ponder my next moves. It was summer 2006 and a time of change. In addition to my divorce, I'd resigned from my teaching and coaching positions. I had cashed out what had been built inside my retirement fund. Teaching is a noble profession, but I didn't envision enjoying it for thirty years. Time to move forward. I'd soon be moving to another town and reinventing myself as a childfree divorcee. The retirement fund cheddar would keep me afloat while my transformation took place.

Everything will be fine, I told myself. Before I relaxed too much, my phone rang.

"Hello."

"Hey man, waz up?" the voice asked.

"Sittin' at the bar," I said. "Who's this?"

"It's ole Wildfire, bitch!"

Ole Wildfire, yes, we have history. He was a solid friend, but we hadn't seen one another in seven years. Following my discharge from the marines, Wildfire and I tore up the nightlife. We ran hard. My last experience with Wildfire had ended with him being dragged off by police while screaming – don't you know who the fuck I am? Not long after the incident, I moved to attend college. During that time, Wildfire's professional life skyrocketed, and he'd also married and divorced.

Wildfire had always been self-employed. During his earlier days, he ran a pressure cleaning business servicing newly constructed apartments in and around Atlanta, Georgia. We referred to his business as a "redneck business"—the type where the tools of the trade can be pulled on a trailer behind a pickup truck. It was blue-collar. Mexican laborers served as employees, and Wildfire even learned to speak Spanish somewhat fluently. It wasn't abnormal to see Wildfire's pickup and trailer parked by the bar while he sat inside, dawning his cowboy hat, pearl-button shirt, and cowboy boots.

Fast forward to 2006. Wildfire, through hard work and family fortune, had upgraded to the part-owner of a successful plastic surgery practice in Atlanta. He partnered with a doctor who possessed impeccable credentials: Emory University graduate, dentist, and oral surgeon who retrained to become a plastic surgeon. Wildfire ran the business and performed laser procedures for which he was certified while the doc did his thing as a surgeon. Wildfire's pickups became BMWs. Pearl-button shirts were replaced by button-downs and tailored sport coats. Starched Wranglers became designer jeans, and cowboy boots morphed into fine Italian leather shoes. The cowboy hat was lost and replaced with a slick hairdo and a few plugs. A metrosexual was created.

"Where you at?" Wildfire asked.

"At a bar in McDonough."

"I'll be down to get you."

It was on.

An hour later, an upscale black Lincoln pulled into the lot. The back door opened, and I hopped in. Wildfire sat to my left with a Coors Light in hand, grinning ear to ear. An older black man drove. He'd already driven over forty miles from one side of Atlanta to the other to snatch me.

"What's goin' on, ole Ronny B?"

"Oh, you know, same ole shit."

"Have a beer," Wildfire said, as he passed a Coors Light. "We ain't gotta worry 'bout shit. We've gotta driver, I know people, and I got 'Mr. Goodtimes' in my pocket."

Wildfire smirked, reached into his pocket, and retrieved a little baggie of blow. I smiled and flashed the horns \m/ with my left hand. It'd been years since I partied like that. We toasted and proceeded to catch up.

"I'm proud of you, man." Wildfire grabbed my neck and yanked us together cheek to cheek. "For going to college and kickin' ass."

"Dude, I'm proud of you!" I replied.

"Yeah, but I'm pissed I didn't finish my . . ."

"Fuck that," I interrupted. "You're fuckin' Wildfire, baby. A businessman and entrepreneur. Wooooo!"

We snorted a bump and proceeded to Buckhead, a swanky district in Atlanta with a popular bar strip. The crowds were large and rowdy. Wildfire and I snorted bumps, drank glasses of fine bourbon, and dazzled the young ladies with our charm. Wildfire was edgy and a pain in the ass sometimes. He'd lunge towards an unsuspecting stranger and blow his burp into their face. They'd startle, and he'd look in the other direction as if nothing happened. Sometimes he'd shout, "There she is," and point straight at a beautiful lady passing by. When she stopped to acknowledge him, he'd adjust his finger to suggest he was pointing elsewhere, and she was mistaken. Sometimes it sparked a conversation for a hookup; other times it drew a "fuck off." I was by his side to say, "C'mon Wildfire," and quell the drama. The best-case scenario, for less confrontation, was when Wildfire crossed paths with a Hispanic male. He loved practicing his Spanish. It kept him occupied long enough for the bar worker to finish taking out trash or stocking the bar. I laughed hearing the drunken mix of cocaine-induced Southern drawl and Español.

Wildfire called his driver to scoop us up and take us somewhere else. As we waited outside, he spotted a black Ford across the street. Giddy from the blow, we sprinted toward the vehicle and jumped inside.

"Oh shit!" Wildfire said.

"What the fuck?" the African American driver yelled.

We exited quickly because we'd unwittingly stormed an unmarked police car. Before we could flee, the officer said, "Hold on a second."

Wildfire leaped into action. "I'm sorry, sir. I thought you were my driver."

Before the officer pried further, I chimed in, "Sir, do you think we'd jump into a police car on purpose?"

The officer couldn't choke down his laughter. "I've been doin' this for twenty years and never had this happen."

Everyone laughed. The cop left and Wildfire patted his leg pocket and said, "Good thing he's cool."

I had moved back to an old, familiar place after my divorce. Kennesaw, Georgia, located twenty miles northwest of Atlanta, is where I attended college. It was an improvement from the depressed southside of town where I was raised. Kennesaw residents are a mix of yuppies, college students, and old skool rednecks who keep southern heritage alive. I hung with Wildfire more often since he lived in Smyrna, a few miles south of Kennesaw. He had the ultimate bachelor's pad—two stories of condo space inside an upscale district with bars, restaurants, and high-dollar shopping fronts. Beautiful women suffused the area. Located steps from his front door was a popular bar. Some weekends the city blocked off the district and hosted parties with rows of beer tents, live music, and beautiful ladies—all under Wildfire's nose.

Wildfire spent a lot of time with Emmy, a slick-talking carpetbagger from upstate New York. When dressed in a designer button-down, Emmy told the affluent ladies he was a "commodities broker" to sell himself. He was, but his occupation got dirty, and dealt with rough customers. We all told bullshit that showed our lives in the most positive light. But Emmy was the master. His industry was booming due to Atlanta's explosive growth. Profits were high.

Our friend circle was small, but acquaintances were limitless. We spent many nights inside the bar located steps from the condo. Each time, a dozen strangers accompanied us back to Wildfire's for the afterparty. Booze, blow, music, and other activities took place until sunrise. Wildfire's home appeared to be the site of an endless party. There were holes in kitchen walls where Wildfire tossed knives like darts. Empty beer bottles littered the corner pool table and cocaine dust sprinkled countertops. An industrial-sized trashcan on wheels rested by a wall. Dresser drawers inside the master bedroom were opened, and clothes were tossed everywhere. On the nightstand, a half-full pitcher of beer sat with several cigarette butts inside. Dead plants and rotted trees served as décor. The most amusing flaw was the holes inside the pool table, caused when a drunken Wildfire accidentally discharged his pistol. He lived a honky-tonk lifestyle.

Three others were close associates to Wildfire: Kenny, Lauren, and Wayne. Kenny was a laidback black dude who was quiet and cool by all accounts. He worked in a restaurant inside the development and supplied Wildfire with blow and an occasional ride from a club. Lauren, twenty-two and a Kennesaw State student, was a typical college girl who enjoyed the nightlife. Despite possessing maturity beyond her years, she was a sweetheart. We bonded over Kennesaw State since that was my alma mater. Ole Wayne was a dude I'd known forever. We shared a birthday and attended the same high school. He'd fallen on hard times, though. Wildfire moved Wayne in since he'd been evicted from his apartment – a property owned by Wildfire's uncle. Wayne was large, standing 6'4" and weighing close to 300lbs. He was balding on top, but rather than shaving his head, he grew his side hair into a skullet. *Why shave it when you can rock it?* He wore a trucker-style mesh cap over his skullet, a handlebar mustache, and a Winston Red was nestled between his lips on the reg.

We're ATLiens. Atlanta rap duo Outkast coined the term in their 1996 album, and locals ran with it. I was born in Georgia Baptist Hospital, downtown on the Boulevard, and raised on the southside, the area is located below Hartsfield-Jackson International Airport and infamously known as The ClayCo.

Classy ATLiens hang in Atlanta's finest places. The crew piled inside Wildfire's Six Series BMW convertible and headed downtown. Wildfire is a well-connected ATLien, too. He knew the person working the door of every club. If a joint wasn't poppin', we left. Never paying a dime to enter, Wildfire would strut to the front and say, "I'm Wildfire, baby," and we'd follow him inside. Some clubs had a hundred-dollar cover, but we were still welcomed free of charge. If there was a line, we'd walk around it like VIPs.

With Wildfire, Emmy and I acted like our favorite 1980's professional wrestling, aka rasslin', heroes. Ric Flair was our number one. Before leaving, we would watch YouTube videos of Ric Flair's promos to get fired up. In fact, Wildfire's nickname derived from old skool ATLien rasslin' legend Tommy "Wildfire" Rich. Rich was a local superstar known for living fast and hard. They say hanging with Rich during his prime in Atlanta was like hanging with Elvis. Ladies loved him. He had a four-day reign as the NWA World Champion but never reached his full potential. His prime was short-lived, and he spent many years performing inside high school gymnasiums on the unflattering indie circuit. His life mirrors Mickey Rourke's character in *The Wrestler*.

Wildfire did a Ric Flair impersonation. "I can't help it. I's born with a golden spoon in my mouth, honey. Ain't my fault I'm dressed in custom clothes, gotta Rolex on my wrist. Wooo!"

When the ladies paid no mind, Wildfire directed his attention towards a dude wearing a Kenny Chesney shirt. He loved issuing backhanded insults to strangers he perceived as beta males.

"The shoes on my feet cost more than that boy made last month," Wildfire said, in the form of a rasslin' promo.

"Huh."

"Yeah, I'm talkin' to you, Chesney boy."

"What the fuck?"

"Son, I wouldn't go see Kenny Chesney if he were playin' my backyard."

Beta wasn't amused by Wildfire's antics, but he didn't flex with Emmy and me standing around.

"C'mon man, let's go," I said, and chuckled. "Why do you do that shit?"

On the way out, Wildfire snatched a tube from the server pacing around with shots, guzzled it, and tossed a twenty on the girl's tray. "Keep the change, darlin'."

Another time, Wayne was outside smoking when we gathered him to head out. A middle-aged man sporting a clean, dark beard strutted by.

Wayne screamed, cigarette dangling, "Oh my God, it's Kenny Loggins!"

Laughter erupted from onlookers as the Loggins lookalike blushed and passed by. Ole Wayne sang, "I'm alright. Nobody worry 'bout me . . ."

I decided to lay low one night because Wildfire's honky-tonk pace was hard to keep up. Despite feeling fresh the next morning, I was unnerved by an eerie call.

"Hey man, you ain't gonna believe this," Wildfire said.

"Waz up?"

"You know Kenny and Lauren, who hang here?"

"Yep."

"Kenny fucking killed her!"

"What?" I asked, while my heart sank.

"Then he killed himself!"

"You've gotta be shittin' me."

Wildfire got wind of this before it was reported in the media. His voice quivered while he spoke. He was shaken because he was possibly the last person to see Lauren and Kenny alive. Wildfire wasn't involved with the murder, but I reminded him an investigator might knock on his door.

"Don't panic, man," I said. "Just clean the fucking condo."

"Why would someone come here?"

"Gee I dunno, Wildfire. There are several calls and texts linking you to both of 'em. I bet you're one of the last texts and calls on both phones."

"Yeah, you're right," he acknowledged.

"Get your shit together."

Wildfire dialed his trusted amigo, Miguel, to come to his condo and earn some overtime cheddar. He'd been with Wildfire since the redneck business days, had keys to the home, and came and went as he pleased. The casa received a makeover. Investigators never showed.

That evening I was glued to the local news. Kenny and Lauren were the top story. Murder in Atlanta isn't always top story worthy; it's a day in the city. But the background and circumstances behind this murder/suicide catapulted it to top story status internationally. A picture of Kenny wearing shiny marine corps dress blues was shown as the story was told.

What the fuck?

Kenny knew I was a former marine and I couldn't help but recall a conversation where he had picked my brain.

"You were in marines?" He had asked.

"Yep."

"Where were you stationed?"

"First year Okinawa, and rest of term at Camp Pendleton."

"That's cool, dawg."

Kenny never mentioned being a former jarhead for good reason. *Holy shit, he's that fucking guy?*

Labor Day 1995, I was stationed in Okinawa and had just turned nineteen. Okinawa is a tropical island located four hundred miles south of the Japanese mainland. Tiny in size, it's home to seven marine bases and over a dozen US military installations. These aren't typical stateside bases that look like a city. They're small camps. Mine was Camp Hansen. A short walk from the barracks door could land you inside a nightclub. No drinking age laws existed to stop young marines from indulging, either. It was wild.

During morning formation, following the holiday, the company first sergeant made a shocking announcement. Two marines and a sailor kidnapped a twelve-year-old Okinawan girl. They bound her hands, duct-taped her eyes and mouth, and beat her. They gang-raped this poor child and left her for dead on the side of the road. The culprits served in our battalion.

This sparked an international incident that strained relations between Japan and the US, leading to the largest anti-American protests on the island since the 1960 Treaty of Mutual Co-operation and Security was signed. The treaty enables the US to place troops on the island, at the expense of the Japanese and prevents Japan's remilitarization after World War II. Chants of "Yankee go home . . . Yankee go home," could be heard during morning formation each day, and I couldn't blame the locals for their outrage. The commander of the US Pacific Command, Admiral Richard C. Macke, was relieved following the incident. He was reduced in rank and forced to retire after this boneheaded statement in a press conference,

"I think it was stupid. I've said several times: for the price they paid to rent the car, they could've had a prostitute."

The three dudes were handed over to Japanese authorities by the US military. They were sentenced to several years in Japanese prison along with hard labor. The corps hit them with dishonorable discharges, too. Each was released in 2003.

Kenny was one of those dudes.

It's hard to believe I rubbed elbows with that piece of shit. Kenny ended up in Atlanta after serving his time. When he struck again, Lauren was his victim. She was beaten, sexually assaulted, and strangled to death inside her own apartment. Hours later, Kenny cut himself in the arms, ripping veins at his elbows, and soon he was dead beside her. The legacy he leaves is evil – pedo, rape, murder, suicide. A twelve-year-old girl's world was forever changed, and a twenty-two-year-old Kennesaw State student's life was ended. Fly high sweet Lauren.

Months later, I agreed to move to Rome, Georgia, at the request of my employer. The Atlanta nightlife was hectic, and an excuse to move is always good by me. I like seeing different places; I'm that guy. Unfortunately, Rome wasn't the place for me, but I tried making the best of it. I rented a cool loft apartment inside Rome's tiny downtown. The town's best bars and restaurants were steps away. Guess you could say I was a poor man's Wildfire.

I was miserable and couldn't stand Rome – neither the people, the mentality nor the layout of the land. The muddy shit river flowing through wasn't aesthetically pleasing. The locals were rough as fuck, wearing mesh caps and dirty clothes. Their hair was unkempt, and most didn't have a full set of teeth. They appeared "rode hard and put away wet". Uneducated is how I'd describe the sound of their dialect. As far as the ladies go, most who were over eighteen were either married, had kids, or both. Being raised on Atlanta's south side, I was no stranger to white trash. But in Rome, it seemed to be the entire population with a touch of ghetto sprinkled in. Narcotics crippled the community. I'm no saint when it comes to drugs, but I have limits. Boundaries didn't exist in Rome. I found myself amongst people who pumped needles and ate OxyContin like candy. Coworkers walked into bathrooms, spent fifteen minutes inside, and stumbled out like zombies. I had a name for these people: The Romans.

It didn't take long for trouble to brew as my frustration mounted. One morning I found myself inside a Roman drunk tank, roughed up, after being sprayed with pepper spray and tased. The night before went sour under the lights of Rome's unflattering bar scene. I had signed up for karaoke inside a crowded redneck dive. But instead of singing Hank Jr, I went on a tirade.

"You fuckin' piece of shit Romans are nuttin' but white trash! I bet y'all have more pills in your pockets than teeth in your mouths. You're a disgrace to Georgia. Fuck you, you fuckin' Romans!"

I dropped the mic and exited the joint while the Roman Po Po waited outside. They attempted their arrest steps from my apartment, but I was uncooperative. The pepper spray and taser were deployed, which prompted me to do the funky chicken. I blacked out and was chauffeured to the clink. In the morning, I called in sick to work from jail via a collect call.

I didn't expect this announcement when my boss answered. "You have a collect call from, Ronny, an inmate in the Floyd County Jail. Do you accept charges?"

The incident resulted in a misdemeanor and a small fine. No biggie. But the worst of my problems was getting fired from my job. A domino effect of troubles kicked off: money problems, taking on debt, and underemployment. What I didn't know is this was a blessing in disguise. The dominoes began to fall in Uncle Sam's direction. This ATLien wasn't compatible with the Romans, so I split.

I spent more time with my father. Mom had died five years before of cancer, so dad welcomed my companionship. At over sixty, he was decaying from the years of blue-collar labor and, we'd later learn, illnesses due to his exposure to Agent Orange during the Vietnam War. He was fighting the VA for disability compensation for a host of issues. Big Ronny B was a Vietnam veteran, yet he was neither a flag-waiver nor one of those guys who wore "thank me for my service gear." He wasn't ashamed of his service, but he never boasted over it.

He once said. "You couldn't give me a million dollars to do it again, but you also couldn't give me a million dollars not to have done it."

My favorite quote of his was from when Big Ronny B was asked to reenlist and offered a squad leader's position. Yet he'd be required to stay in Vietnam for a second tour.

The first sergeant had asked, "Son, what'd you do for a living before getting drafted?"

Big Ronny B replied with his blue-collar vocation.

The first sergeant mocked it and asked, "Is that what you wanna do for the rest of your life?"

Big Ronny B, cool as ice, clapped back. "First Sergeant Hernandez, I'd rather take the backseat outta my car and haul cow shit for a living than stay in this motherfucker and get shot!"

My father was my hero. I adored him. He and my mother managed to keep our family afloat during trying times. He twice suffered debilitating injuries at work that led to lost time and lengthy recoveries. One time, he fell over twenty feet onto a concrete floor. The other, he almost cut off his finger in a saw and severely jostled his neck. He'd suffered through his share of surgeries, physical therapy, and court litigations. But he returned to work after each injury. No matter what he went through, he managed to be a positive person. He possessed the gift of the gab too. He'd say, "Yes ma'am," to the sixteen-year-old behind a cash register since he wasn't a salty boomer.

And if you thought you could diss ole Big, he'd chop you up with a haymaker-style comeback.

I once asked, "Why don't you blow me?"

"I'll blow behind your ears while you're on your knees."

I was disappointing my father though. He knew my potential, and it hurt that he witnessed my latest failures. But they were merely setbacks. Following my divorce, I asked myself: *Why not return to the military?* Two wars were going on and I was able-bodied. I still thirsted for the combat that I never received as a jarhead. Now I was available. No excuses. My father wasn't crazy about the idea, but he understood. He knew Atlanta no longer had anything good to offer me. So, I stood before Sam and raised my right hand, again.

Wildfire supported my decision and even encouraged it. A proud veteran of the air force, he had taken part in Operations Provide Promise in Panama and Provide Hope in Turkey. Wildfire often reflected on his time in the service, especially after a few drinks, and he'd speak about how he missed it. He envied me for being able to return.

CHAPTER 2: BACK IN THE SADDLE

"I don't give a fuck what service you came from or what rank they gave you. Who the fuck do you think you are? Here you're a freaking private training to be an infantryman!"

That was the informal greeting blasted to Willie B and me as we were welcomed to First Platoon, Echo Company, of the 2-19 Infantry Training Battalion in Fort Benning, Georgia.

Willie B and I joined the young men of First Platoon before the crack of dawn and commenced getting smoked. We performed mountain climbers and side-straddle-hops in a field. The humidity was oppressive. Following our warmup, the platoon assaulted a small obstacle course. We climbed ropes, scaled walls, swung from monkey bars, and crawled through dirt while shouting motivational chants. The sun hadn't pierced through the clouds yet, but here I was grunting through an intense physical training session alongside men fourteen years my junior.

"Left, left, left, right,"

"SHOOT, MOVE, COMMUNICATE, KILL!" the platoon shouted as the drill sergeant announced the right foot in the cadence.

"Left, right,"

"SHOOT, MOVE, COMMUNICATE, KILL!"

"Sitting in a foxhole . . . sharpening my kni-ife . . . in walks the enemy . . . I had to take his li-ife . . . airbo-oor-oor-orne . . ."

As we marched to the DFAC (Dining Facility) for morning chow, we shot, moved, communicated, and killed. We slashed the imaginary enemy's throat and stabbed his chest with a fighting knife. We led the way on the battlefield, assaulted from the sky, and laid waste to everything. We were airborne infantry. While standing in line to enter the DFAC we shouted the Infantryman's Creed.

"I AM THE INFANTRY. I AM MY COUNTRY'S STRENGTH IN WAR, HER DETERRENT IN PEACE. I AM THE HEART OF THE FIGHT—WHEREVER, WHENEVER. I CARRY AMERICA'S FAITH AND HONOR AGAINST HER ENEMIES. I AM THE QUEEN OF BATTLE . . ."

Following a southern breakfast complete with scrambled eggs, grits, and sausage patties, we layered in full combat gear and headed to the range for a stress shoot. The sun sizzled. I tossed the loaded rucksack over my shoulders. It added grueling pounds to the helmet, armor, weapon, and ammunition already weighing me down. "Kill" was screamed, and we sprinted off on a mile run. Sweat dripped from my pores, drenching my uniform. I drove through my discomfort while my boots pounded the steaming pavement. After completing the run, we leaped into a prone shooting position and engaged targets. The point of this exercise is to shoot with precision while fatigued. During this week of training, we completed a five-mile run, a three-mile run wearing full combat gear, and a ten-mile road march with a full rucksack, a weapon, and armor. We had to drink as much water as possible because the West Georgia heat zaps your energy. One young man was sent home because he passed out during the march. His core body temperature rose so high that it almost cost him his life.

At thirty-two, I was grinding in infantry school with the young bloods. It was obvious; I was the oldest trainee in the platoon. Questions came my way from privates, cadre, and even drill sergeants:

"What are you doing here?"

"Why aren't you at OCS?" (Officer Candidates School)

"Do you have a death wish?"

I'd answer, "I just wanna be here," and smile.

I kept the past buried and moved on. Some asked—why I didn't return to the marines? I had approached them, but they weren't interested once I revealed my age. Marines love young fresh meat. They're the smallest branch and can be picky. I wanted something different, anyway, and the army needed me. All I asked for was the E-4 stripes I earned during my first enlistment. Never in my wildest dreams did I believe I'd be subjected to another dose of boot camp, though. I smuggled in ibuprofen to help my body cope with the day-to-day grind. Thirty-two is considered prime in civilian life, but here I might as well have been a senior citizen. I was older than both platoon drill sergeants – Senior Drill Sergeant Paske and Drill Sergeant Weaver. I likened my situation to that of a thirty-two-year-old professional football player. At that age, the footballer is in the twilight of his career. I knew I could perform any task these young bucks could. But I had to be smart.

Before arriving at infantry school, I was required to attend a five-week refresher at Fort Sill, Oklahoma, called the Warrior Transition Course (WTC). It was designed for prior service members who joined the army. If someone separated from the navy or air force and wanted to continue their military career in the army, they had to complete this training. If a former soldier or marine hadn't been on active duty for over four years, they also had to do this bullshit. One stark contrast between infantry school and WTC was the age of the trainees. Most guys at infantry school were recent high school graduates. WTC trainees ranged in age from guys in their mid-twenties to one gentleman who was fifty and had been out of the navy for twenty-five years. Most dudes at WTC were older than me. There was nothing grueling about WTC, yet we often returned from our morning jog to a tossed barracks. It looked like a hurricane had blasted through, ripped bunks, and tossed them to the floor. The garbage from knocked-over trashcans mixed with uniforms from some douche bags' unsecured wall lockers. WTC drill sergeants loved fucking with us for the sake of it, too. Most of us felt it was unnecessary. Why treat older prior service soldiers as privates who'd never served a day? There were former rangers, paratroopers, recon, army, and marine corps grunts in the class.

The economic realities of the US were on full display at WTC. Many guys were laid off from previous jobs, struggling to find work, or having trouble supporting their families. The economic collapse was great for the army, though. America was knee-deep in a two-front "War on Terrorism" in Iraq and Afghanistan. The army was stretched thin. Inactive soldiers were called to active duty. Soldiers who planned to leave the service at the end of their terms were "stop-lossed" to perform another deployment. Twelve-month combat deployments were extended to fifteen months. Many soldiers were serving multiple combat deployments with minimal dwell time between tours. It was the summer of 2008, and the army needed more asses. WTC did its part to fill the ranks with saltier men.

I met Willie B at WTC, and we hit it off. He was a twenty-two-year-old former airman from Vermont. I teased him for being mighty white due to his glasses, fine speech, and polite demeanor. It falsely led one to believe he lacked swagger. Playing off that idea, I altered his first name "Will" and his last name, beginning with B, to come up with his nickname. It added flair. The original Willie B was a famous silverback gorilla that lived inside Atlanta's zoo for almost forty years. The gorilla was named after William B. Hartsfield, one of the last white mayors of Atlanta, who served until 1962.

Willie B and I were disappointed about being dropped into a basic training platoon following WTC. Word was we'd pick up training during AIT (Advanced Individual Training). Instead, we were in basic training during week four. *Why did they even send us to WTC?* I also learned there's no AIT in infantry school; it's four months of boot camp. We were also told we'd receive special privileges at Benning for prior service. That was laughable.

"What the fuck is wrong with you, bitch?" Weaver screamed.

No one knew who he was screaming at since we were locked at attention. "Yeah, I'm talkin' to you, four-eyes!"

That smoking gun indicated he was addressing Willie B. I laughed inside.

"Yeah, shit stain, scratch your motherfucking face again in formation." The drill sergeant went on.

"Fuck this bullshit," Willie B said later. "In the air force, I was a general's aid with an office inside the pentagon."

"Now you're here with a teacher, and we're gettin' the biz."

We chuckled and continued bitching.

"Out of sight, out of mind" was my philosophy when I was a young buck on Parris Island. This approach worked well at WTC, too. I blended in amongst the crowd and blew through training unnoticed. I had no desire to be the hero there. But I stuck out like a sore thumb at infantry school. It was my age, arriving late, and guys knowing I was a jarhead. I still managed to fit in. The privates respected that I performed all the training side-by-side with them. I was also not above taking part in some mischief. Such things are prevalent when college-aged men are housed together for an extended time. I marveled over watching these guys progress from kids to military men and develop discipline, toughness, and loyalty.

I was fond of one young man in the platoon, a kid from Los Angeles named Joey. Lighthearted with a great sense of humor, he stood taller than average, had blond hair, blue eyes, and large feet. A little nerdy mixed with California cool, Joey C was a film buff (some of his family members were involved in the film business). He recited movie lines while impersonating voices. Joey also "dug" professional wrestling, which made us closer. He'd heard of most of the old skool wrestlers whom I'd long admired. He was even familiar with Tommy "Wildfire" Rich.

One morning as we were getting up, Drill Sergeant Weaver approached the platoon standing at attention and asked, "What was Motley Crue's first album?"

I belted out. "Too Fast for Love, Drill Sergeant!"

Weaver cracked a smile.

"You old motherfucker."

He later called me into his office. "Look, Bruce, I know you've already been here and done this shit. You're doing a good job. So, if you show up on time, in the right uniform, and continue to set a good example, no one is gonna fuck with you."

Weaver paused and continued. "You understand we've got two wars going on, and these kids, and you, are guaranteed to see combat. Understand there's a method to my madness."

"Understood, drill sergeant." I intoned.

"You've played this game before. Now get the fuck outta here, jarhead."

Drill Sergeant Weaver did his best to prepare us for the horror that awaited. It was on him to ensure eighteen-year-old kids possessed the mental and physical toughness required for war. Plus, a drill sergeant wouldn't be serving his men well if he wasn't playing the part of a slave-driving, demanding motherfucker. Weaver had served several infantry tours in Iraq and knew the enemy didn't give a fuck about our feelings.

Weaver always had our backs, though. While released for a two-hour pass, an off-duty drill sergeant decided he would impress a gal. He was driving a sports car, which rolled by as we walked to the PX. The driver shouted, "Hey, get the fuck outta the road!"

"Fuck you!" I yelled back.

The vehicle stopped alongside the curb, and the sergeant's girl screamed. "Who the fuck you think you are?"

The drill sergeant, wearing civilian clothes, exited the car, stepped into my grill, and demanded. "Stand at parade rest!"

"I'm not standing at parade rest for you." I countered.

"The fuck you ain't."

"Lemme see your ID card?" I asked.

"I'm not showing you my ID!"

"Well, if you don't ID yourself, I'm not standing at parade rest, and I damn sure ain't listening to any more of your bullshit." I walked off.

During the episode, the gal sitting in the car yelled. "Get his name!"

I was later called into Drill Sergeant Weaver's office. "Tell me what the fuck happened, Bruce."

I spat the story and added. "Drill Sergeant, he never presented identification, so in my mind, he could've been some enlisted guy, maybe even a civilian."

I met his gaze and continued. "He could've been anybody. I thought he was some Joe trying to impress his girl."

After corroborating my story with others, Drill Sergeant Weaver reported his opinions to the first sergeant. I was praised for my handling of the situation despite this dude indeed being an off-duty drill sergeant. The drill sergeant was reprimanded by his command.

While we scrubbed shitters one evening, Weaver announced. "The platoon bulldog took care of biz."

Looking back, it was a blessing to bear the brunt of training. It got me into peak physical conditioning again, and it served me well when I moved on to my permanent assignment. In a "turning blue" ceremony at the end of training, the men were pinned with the infantry blue cord, a fourragere. It's worn around the right shoulder of dress uniforms. It distinguishes infantrymen from all other soldiers.

Drill Sergeant Weaver pinned the blue cord around my shoulder. "Congratulations, Bruce; you're one of the finest soldiers I've trained."

I shook his hand. "Thank you, Drill Sergeant; that means a lot."

Before moving on, Weaver looked into my eyes coldly. "You're gonna be a big "DICK" in whatever unit you go to. Guys will look up to you. I know you won't let 'em down."

He wasn't calling me a dickhead or an asshole; he was using a term blasted at privates on Fort Benning for motivation. A DICK was a Dedicated Infantry Combat Killer.

Willie B was headed to airborne school, and I was handed orders to Alaska. It was fine; I wanted as far away from Georgia as possible. Plus, my buddy Joey C was headed there with me. Uncle Sam was helping me hit the reset button.

CHAPTER 3: NORTH TO ALASKA

The cabbie drove from Ted Stevens International Airport in Anchorage to nearby Fort Richardson. Everything was covered with blinding snow. It wasn't the first time I'd seen snow—maybe the third or fourth. During the previous times, everything shut down from a dusting. Anchorage was receiving a foot today, and all was still alive and thriving. People walked, jogged, drove, and carried out their normal routines as if everything was all good. The sunlight beaming off the surrounding snowcapped mountains was mesmerizing.

Joey C and I met and settled into our new environment. There's anxiety when checking into a new unit, which is normal. I knew from my marine corps experience how "boots" were treated. Everyone fucks with boots. In the army, new guys are called "cherries." Whether army or marines, no one gives a fuck about new guys. They're unproven. During peacetime, new guys proved themselves through indoctrinations or drunken bravado. I proved myself back then by fighting and lost two week's pay. That wasn't my style anymore. Besides, establishing yourself in an army that'd been at war for over seven years involved combat cred. Flexing my marine corps muscles wouldn't play here. *No big D.* Checking into a grunt unit would be fun. *At least I'm no longer a POG* (Person Other than Grunt).

"Hey man, you notice everyone's wearing an airborne tab?" Joey C asked while we chilled inside a waiting room.

"Yep."

"We ain't airborne."

"No, the fuck we ain't," I said, casing the dull room.

"What's up with that?"

"I dunno, but we're about to find out."

Uncle Sam can flip the script; I wasn't surprised since this wasn't my first rodeo. Indeed, Joey C and I were checking into an airborne unit despite not being airborne qualified. This gave us another reason to be fucked with. Airborne guys refer to non-airborne as LEGS (Low Energy Ground). However, it's not normal for LEGS to serve in an airborne unit. Weaver offered me an airborne contract before leaving Fort Benning, but I'd already seen my orders to Alaska. No unit or other information was given. I chose Alaska over airborne. *Who knew?*

As the room filled with "cherries," thirty or so, and none of us airborne qualified, I figured it out. I passed it on to Joey C.

"Dude, this unit is about to deploy," I whispered. "And we're goin' with 'em."

"What makes you say that, man?"

"Why else would this many LEGS be checking in here?"

Joey C was young but not stupid.

"They need more swingin' dicks," I said. "We're it"

We were checking into the Fourth Brigade (Airborne), Twenty-Fifth Infantry Division (4/25) and passed to the Third Battalion, Five Hundred Ninth Parachute Infantry Regiment (3/509). The 25th Infantry Division is headquartered at Schofield Barracks in Hawaii. The 4th Brigade was a new unit that activated in Alaska in 2005. It's the first airborne unit created inside the US since World War II and is the lone airborne brigade in the Pacific Theatre. The 509th Parachute Infantry Regiment, Geronimo, performed our nation's first combat jump on November 8, 1942, as part of Operation Torch in North Africa. It followed the longest combat invasion in history – a sixteen-hundred-mile flight from England. The 509th participated in four combat jumps during World War II; two were conducted in North Africa, one in Italy, and one inside France. Those are huge boots to fill.

The brigade was set to deploy to Afghanistan in March 2009. We had four months to enjoy Anchorage before taking to the battlefield. Uncle Sam did throw Joey C and me a bone, thankfully. The brigade had just left for the National Training Center (NTC) at Fort Irwin, California, to train. They'd be gone a month, and the cherries had little responsibility and not much to do.

I explored Anchorage. Joey C and I rented a van and some hotel rooms. Time to party. I asked the locals what was hot, and 4th Street was the place. Small bars lined the street, side-by-side from one another. Since Joey C was underage, he dropped me off inside the bar district and went to hang out at a joint catering to minors.

I bounced from the Pioneer Bar to the Gaslight Lounge and took a ride to Chilkoot Charlie's, aka Koots. Koots was the place to be. It was unique—with its sawdust floors, tree stump barstools, and wooden bars, and it attracted a diverse crowd. Different rooms housed different parties, and a live rock band played one of the rooms. Feeling at home, I took to calling the local sexy young ladies "snow bunnies" for obvious reasons. Their furry boots were hot. *There's a line of sexy women's clothing for the harshest winter climates. Who knew?*

My self-esteem was building, and I felt good about my path. Yes, I was a nobody in the military, but I had a purpose. Had I remained on active duty since 1994, my rank would have been much higher. But I was back in the game and had risen out of my funk. My body was solid, and razor-cut even on my shorter-than-average frame. Ten thousand dollars was stacked in my account, and the twelve-month deployment would allow me to save an entire year's salary. There wouldn't be much to spend money on, and hazard pay would be added while federal tax deductions ceased. It was a perfect opportunity to create a nest egg.

When I hit the town, I ensured I was looking "fly." I'd tell the younger guys. "You've gotta wrap the package nice every time you leave the house."

I'm a bit metro, too, so I dressed in tune with proper men's fashion, practiced good hygiene, and employed proper manscaping techniques. I wasn't preppy. There was no golf shirt tucked inside khaki slacks with loafers on my feet nor a sweater tied around my neck. But I dug fresh jeans, leather shoes, and a tailored sports coat during the wintertime. I borrowed from the ole Wildfire and Emmy playbook. Sometimes I rocked a pearl button shirt with a bit of my rock-hard chest exposed. An untucked pearl button mixes country/western with metal, and I added a touch of Ronny B class to smooth over my rugged edges. I'd give the boys advice.

"Tennis shoes are for the gym."

"Lose the hoodie and saggin' jeans and wear clothes that fit."

One trait that made me stand out amongst the Alaskan crowd was my southern accent. Many women outside the South have a thing for the southern dialect. They melt when they hear it spoken the proper way. I spit a smooth, eloquent Georgia twang that exudes calm and puts people at ease.

After I met a lady, they'd ask me. "Where are you from?"

"I'm from Georgia, darlin'," I'd reply. Or "Atlanta, GA, sweetie," saying the letters G and A separately.

My southern charm and swagger made me popular with the Alaskan females. Good looks and prime manhood didn't hurt, either. I had a thing for those "sweet young thangs" who roamed about in their sexy snow bunny boots. An ATLien had landed. *I'm not your average Carhartt-wearing Alaskan male.*

The Ric Flair lines and 1980's rasslin' swag were part of my arsenal. Sometimes during a conversation, I'd ask, "It's nice to be talkin' to a real man, ain't it, dear?"

If an interested lady reacted "chippie," I might say, "You just be glad I'm here right now lookin' the way I look, talkin' the way I talk."

The ladies knew it was sarcasm. But they understood I had confidence in myself, and I was a man. A boy, a weak pushover man, will appear uncertain and needy. He'll lack confidence in himself and his actions. On the other hand, a real man takes control of a situation and displays leadership. A good woman will follow his lead. At least that's what I've learned. I could be wrong, though. Southernisms such as: "sweetie," "darlin'," "honey," "sugar," "dear," and the like are proper greetings to us Southern gentlemen. We pass them on to the ladies to display our respect and generosity. It's not always a term of endearment.

I spent quality time around Anchorage with a gal named Jane. We met in a funny way. While I was hanging at the Peanut Farm sports bar with friends, a dark-haired, fair-skinned beauty pranced to the seat at the bar beside me and made herself comfortable. She wore those oh-so-sexy knee-high snow bunny boots and had a black pea coat draped over her tight body. *This is the whitest dark-haired gal I've ever seen. This is Alaska.*

A World Series of Poker episode was playing on the television. One player was dealt a pair of Kings while another player called his raise with a shitty hand.

"You know, I bet the guy with the garbage hand wins," I said to her as my opener.

"Oh yeah?" she asked. "That one guy has a pair of Kings."

"Yeah, but I bet the other dude draws out on him."

"I don't think so."

I assumed the Kings would prevail, so I tossed out an idea. "I'll tell you what, if the Kings win, I'll buy you a drink."

The pair of Kings won.

"I'll be damned. I guess I owe you a drink. What would you like, dear?"

After returning with her drink, I introduced myself and asked her name, adding a southern sweetie to the mix.

"Jane," she said.

"Ole sweet Jane."

"Where are you from?"

We started to converse, but I could see Jane was a bit reluctant.

"I'm as gentle as a lamb and sweet as candy, darlin'." I reassured her.

My positive intentions and witty charm won the night. I told the boys. "See y'all later?"

My weekend routine started with staying at the Marriott downtown to get away from the base. The military rate was under a hundred bucks per night, and I had no problem paying it. During the day, I'd rest and recover. At night, I'd peep the bar scene. In a few months, I'd be grunting inside Afghanistan for a year. *Why not live?* Reality stared into my hazel eyes. *I may not make it back alive.* Big Ronny B often said – "live for the day, fuck tomorrow." I bet it was the Vietnam in him, knowing tomorrow may never come. So, I lived for the day.

Jane arranged to hang out at the Marriott one weekend, so I booked a room up high with a sweet view of the city. I peeled back the curtains to display the Anchorage skyline and sat an ice bucket containing a bottle of wine on the table in front of the window, alongside a single rose. Jane melted when she arrived. We attended an Aces hockey game inside the Sullivan Arena, shopped the Fifth Avenue Mall, sipped drinks in the surrounding pubs, and enjoyed one another's company. She was only twenty-three, a full-grown woman to Joey C and the other boys. But she was a sweet young thang to ole Ronny B.

Reality settled in when the brigade returned. Cherries were rounded up and sent to battalion headquarters to await assignment to the company level. Twenty or so of us sat inside a tiny room when representatives from the companies arrived. A young sergeant appeared and called Harry, Barrett, Joey C, and me. He informed us that we'd be coming with him to Charlie Company. The four of us were already acquainted from Fort Benning and excited to stay together.

"I don't do all that purdy stuff," the young sergeant said. "So, line up and follow me."

Sergeant Dutch was an Alabaman who possessed a laidback southern swag I was familiar with. His voice resembled Matthew McConaughey's character from *Dazed and Confused.* "Alright, alright, alright," went through my head as we route-step marched to Charlie Company Headquarters. "That's what I like about them high school girls . . ." also crossed my mind. Dutch's uniform said Ranger School and Jumpmaster. I learned there wasn't many of those walking around the brigade. *The dude's the real deal.*

"Alright, alright, two of y'all are going to Second Platoon, and two of y'all are headed to Third," the sergeant said. "How do y'all wanna split?"

Joey C and I volunteered for Third Platoon since we'd been together since Benning. Harry and Barrett chose Second.

"Good luck, gents," Sergeant Dutch said before sending us on our way.

The basement of the Charlie Company building served as a rallying point for three infantry platoons. The men hung out there before formations. After formation, they'd disperse to either train or maybe play video games. A glance inside the latrine (bathroom) gave you a feel for the attitude. Placed inside a urinal was a Jane Fonda sticker captioned "Hanoi Jane" with her face pinpointed as the bullseye. I could feel the burn of the men's eyes when I entered the building for the first time. Cherries were being sized up. There was more testosterone inside that joint than a male prison yard.

Our physical fitness, knowledge, and toughness were tested each day. There wasn't much time to mold cherries, and non-hackers needed to be weeded out. On day one, we got smoked—pushups, flutter kicks, back and forth. Welcome to the real infantry. Our job was studying the Ranger Handbook, becoming efficient with weapons, maintaining superior physical fitness, and displaying more discipline than the regular soldier. My prior service had its advantages, though. I understood the military culture and was rarely fucked with.

The one other time I was smoked was inside the education center. We were studying Pashto, the language spoken by the Pashtun people of Afghanistan. My computer wasn't working, so I sat twiddling my thumbs.

"What's your problem?" Smeed, a specialist (E-4), asked.

"Nothing."

"Then why are you sitting there doing nothing?"

"My computer is fucked up!" I'll admit I was too chippie.

Staff Sergeant Hix walked in with the other squad leaders. Probably the tallest guy in the platoon, slim with sandy blond hair, Hix loved old heavy metal. He also knew how to lay down the law. Hix approached, kneeled, and said, "Get the fuck outside," into my ear.

In the hallway, I did push-ups, side-straddle-hops, mountain climbers, and flutter kicks. It was embarrassing since civilians, dependents, students, college instructors, and library patrons roamed the area.

As I held the push-up position, arms trembling and sweat dripping from my brow, a voice spoke, "Is that how they do it in the marines?" Hix asked. "Tell a team leader to fuck off?"

Oops, Smeed is a team leader?

I met his gaze. "I take orders, sergeant, but I don't take shit."

"Get the fuck up, Bruce."

I believe my response garnered respect from Hix and the other squad leaders. They wanted disciplined and obedient soldiers, for sure. But displaying a backbone scored points with mid-level leadership. Smeed turned out to be a good dude with an even keel. He was intelligent and had lofty aspirations for life outside the army in Wisconsin. In contrast, Hix was a fiery guy from Washington state. He loved old heavy metal and football. Metal guys stick together. \m/

Joey C and I were assigned to Second Squad. It was led by Staff Sergeant Stan, a high achiever who gained rank fast. At around six feet tall, he had dark hair, a medium build, and hailed from Pittsburgh. He wasn't always a prick, but he could be one on the wrong day. There was a method to Stan's madness, though. His aggressiveness stemmed from two things. One, he'd been "stop-lossed" by the unit. Rather than returning to Pittsburgh, he was forced on another grueling deployment. Two, he was leading a squad full of cherries and had three months to get his cherry squad ready for battle. Half of us didn't train with the platoon at NTC, either. Stan was taking a big stiff one with no lube from Uncle Sam. Joey C nicknamed him "Iron Guts" after getting smoked. It stuck.

Our Second Squad was broken into two four-man fireteams: Alpha and Bravo. I was placed in Alpha Team, led by Corporal Stu, and Joey C went to Bravo, where Sergeant Parthia was in charge. Stu was a good ole white boy from Utah who was built like a Greek statue but with sleeves of tattoos. Raised as a Mormon, he didn't seem into it anymore. An eagle scout, he loved the outdoors and was married with two kids. Bravo's leader, Sergeant Parthia, was laid back. He never raised his voice and possessed a good sense of humor. This was shocking since he was a stop-loss guy who'd be spending the year in the suck rather than at home in Maryland. A jokester, he was a good fit to lead Joey C.

The NCOs (Non-Commissioned Officers) in the platoon were on their second deployment at least. The 4th Brigade deployed to Iraq in late 2006, and the unit served an extended fifteen-month tour during Iraq's surge. They were experienced combat veterans, which was assuring.

"I zoo no give e fook bout de Moorines or whatchu zid dere," Manny Fresh, the platoon sergeant for Third Platoon, said, when Stan took Joey C and me to his office to introduce us. "You two zoo ze fook what yer told and no proobl emz."

Manny Fresh was a Venezuelan with a thick accent. He was difficult to understand, especially over the radio. I sensed the squad leaders didn't like him. Stan's mood went to hell following meetings with Manny. Much of the discontent stemmed from issues during the deployment and the past year of training. Squad leaders despised Manny's leadership style. But those issues were between the NCOs and Manny Fresh. I wasn't there to question anyone nor discern who was right or wrong. As I'd been trained, I stayed in my lane.

"Yes, sergeant," we said as we stood at parade rest.

While Manny gave his spiel, I glanced at Stan, who rolled his eyes. When Manny paused, I said, "Yes, sergeant," displaying the respect that he seemed to crave.

When Stan introduced us to the platoon leader, he didn't say much. "Welcome aboard, gents."

The platoon leader, Sammy, was a fresh West Point graduate. He had ginger-colored hair, blue eyes, and a fair complexion. Relaxed yet motivated, Sammy was like many lieutenants – knowledgeable yet inexperienced and likable but at a distance. He held the true power in the platoon. But, as lieutenants do, he yielded his power to Manny Fresh, the salty platoon sergeant. Sammy was at least a second-generation military officer whose father was highly ranked. This Georgian grunt surmised that Sammy aspired to someday be a general. Afghanistan would be his proving grounds.

Unfortunately, a lieutenant looking to make a name for himself could lead his men into unnecessary firefights. Generals love to gloat over combat stories from their time as a young platoon leader.

I couldn't help but think about my military lineage compared to Sammy's family tree of military men. Although I possessed the same level of education, albeit not an elite West Point one, my military lineage came from the dark side of draftees and enlisted men. My father was one of five in his immediate family to get drafted. The fact that my father and his brothers were drafted was a clear indicator of my family not coming from much. There were ways out of the draft and military service, including college deferments. These were options that men from moneyed families could choose. My father and uncles didn't have the means, so they had no choice but to give their lives to Uncle Sam.

Corporal Stu, Alpha Team leader, introduced me to other squad members. Dottie, in the Alpha team, was from West Virginia. Large by infantry standards, he was biracial (white and black) and easy-going. I perked up when he mentioned loving old skool rock and KISS. Dottie owned a sweet-ass Ace Frehley signature guitar that he kept inside the barracks. His favorite KISS song was *Hide Your Heart*—odd since the tune was from the late 1980's non-makeup wearing days. Phife was a Bravo Team guy who was one of the more experienced junior platoon members. He'd been in the platoon a year and earned his Expert Infantry Badge, which required passing a test.

Weeks later, the squad was completed when Blackie and Brandon arrived. Brandon was twenty-three and from Pittsburgh. He loved hunting ducks and whitetail deer. Blackie, from Chicago, had sleeves of tattoos on both arms from his shoulder to his knuckles. A wild and crazy fucker, he was smoked on his first day in Charlie Company for going on a drunken rampage. He was only nineteen, but he stole a club bouncer's car and had a joyride around Anchorage. He blew through red lights, skidded tires, and ended up being pulled over by the police. Instead of getting arrested and charged, the police contacted the unit. Manny picked Blackie up from jail. Manny didn't know his new private because Blackie hadn't checked into the company yet.

"You peez of sheet mooza fooker," Manny shouted, during the ride from jail. "Wut ze fook were you tinkin Vlackie, you fookin poosey!"

Blackie wasn't in danger of being booted, though. Uncle Sam needed all his guys. Some soldiers tested positive for illegal drugs during random urinalysis. They were knocked down a rank rather than kicked out and served "a big chicken dinner"—what marines call a bad conduct discharge.

In the last days of February, Stan gathered the squad to give information about the deployment. "Don't think this shit is gonna be chill. Just because it's not Iraq doesn't mean shit. There's bad stuff going down there."

Stan explained we were headed to Paktika Province in Eastern Afghanistan along the Pakistan border. He gave basic info about the climate, people, culture, geography, etc.

"Get yourselves focused," Stan said before releasing us on our final weekend in Alaska. "Shit's about to get real."

Stan was correct when he said don't believe Afghanistan was chill. At the time, Afghanistan was a forgotten war, far from the average American's mind. Whatever the reason (neglect, the Taliban's hiatus, or assumption the enemy was defeated), Afghanistan was placed on the backburner. In reality, the Taliban was reemerging.

Newly elected President Barack Obama was employing a new strategy focused on Afghanistan. Throughout the years 2009-2010, the plan was to triple the number of US troops to one hundred thousand. The president was surging Afghanistan. What comes with a surge? More troops, more fighting, more injuries, and more deaths. The US was taking its best shot at defeating the Taliban once and for all. That's an ambitious proposition considering Afghanistan has never been conquered—not by Alexander the Great, not by Genghis Khan, not by the Persians, not by the British, and not by the Soviet

Union. The landlocked nation, with unforgiving mountainous terrain, is difficult to navigate. Foreigners have never had a successful prolonged military campaign there. The Paktika Province was one of the poorest areas on the globe and also a Taliban stronghold. Ninety-nine percent of Paktika was considered rural, with little access to running water or electricity. The area's literacy rate was less than twenty percent. Thirty-four US troops lost their lives in the province before our arrival, including famed professional footballer turned Ranger, Pat Tillman. Afghanistan was about to become ingrained into the minds of Americans as the war in Iraq cooled.

A large deployment ceremony for the 4th Brigade was held inside the Sullivan Arena in Anchorage. Governor Sarah Palin was the guest speaker. She spoke about her son serving in Iraq in a motherly tone, and she thanked us for our dedication and sacrifice. She bragged that our unit was making Alaska proud.

Captain Mac and the first sergeant held one final company formation on our last Friday in Alaska. Captain Mac gave a motivational speech to send us off. He said this would be the last time we'd see one another as a company until we arrived back in Alaska. I then realized most of our movements and patrols would be performed in small groups—no larger than a platoon. I looked around at the stone-cold faces in formation and saw everyone was ready to lock and load, starving to kill a motherfucker.

Jane and I shared a final rendezvous under the darkness of the pristine Anchorage skyline. We kissed, held one another, and said our goodbyes. Knowing you're leaving for an entire year makes the moment strange.

"Please be careful." Jane smooched my lips and massaged my neck. "I'm gonna miss you so much."

"I'm gonna miss you too, darlin'," I said as we held each other one last time.

I said goodbye to my father and sister over the phone and made peace with God as I prepared for my journey. The time had come.

CHAPTER 4: WELCOME TO MALEKSHAY

A bird breezed across the mountainous landscape below me as I peeked through the window of the helicopter. "Kalats" (fortified homes made of mud) and "wadis" (dried riverbeds serving as roads) interlaced the villages. Inside the valleys, herds of goats roamed, and crops bloomed. We flew low, and the bird swerved away. The pilots were Russian contractors—yes, descendants of communism, heated Cold War rivals, and prior occupants of Afghanistan.

What the fuck are they doing here?

I grew up when the Soviet Union was vilified as an evil empire. Nuclear war seemed inevitable. I stared into each face of my comrades.

Think any of these dudes remember the Cold War? Most were still shittin' yellow when the Soviet Union collapsed.

The chopper landed on the Helicopter Landing Zone (HLZ), and our boots hit the dirt. Soldiers manning the outpost boarded the helicopter following our departure. Their war was over. Their dust-crusted faces had eyes with thousand-yard stares, and ripped uniforms stained in mud, and blood clung to their bodies. They were thin and shuffled like the living dead into the aircraft. They'd done what we were about to do.

Once we were fully loaded with weapons and combat gear, we hiked to the top of a mountain where COP (Combat Outpost) Malekshay was located. The HLZ was about two hundred meters below the COP, and we struggled to ascend the steep hill. Malekshay stood over 8,000 feet above sea level. To put that into perspective, Denver sits at 5,280 feet, one mile above sea level. Nothing prepares your lungs for the strain of high altitude. You struggle and acclimate slowly.

I couldn't help but feel the thrill; adrenaline overrides fatigue. Joey C stuck out his fist for a fist bump and said, "Here we go, Ronny B."

While taking my first steps in a warzone, I thought about men of war from my family and heritage. I envisioned the iconic Scotsman King Robert the Bruce mounted on a horse, clad in medieval armor, leading a successful charge for Scottish independence. I thought about Thomas Riley Barrett, my great way-back granddaddy, marching in a column across a fierce Civil War battlefield. He met his demise in the Battle of Gettysburg on July 2, 1863. I toasted my great uncle Claude for killing Nazis in Europe as a paratrooper. And I flashed the horns \m/ for my father, who patrolled jungles and rice paddies in 'Nam, keeping an eye out for Vietcong guerillas. Though not yet battle-tested, I was one of these men. We were one. I felt worthy of the blood flowing through my veins, worthy of my country.

I'm a fucking warrior.

The tininess of COP Malekshay was stunning; it was half the size of a football field and shaped like a triangle. Three guard towers were located at the triangle's tips. Hesco barrier walls filled with dirt connected the guard towers. Barbed wire surrounded everything. Guard towers were named by direction: North, South, and East. Below the North Tower, a makeshift gate was the only way in and out. A curvy gravel road descended from the COP. Humvees with Mark 19 automatic grenade launchers mounted to turrets parked beside each tower. Inside each tower were two crew-served weapons: an M240 Bravo and a .50 caliber machinegun.

Malekshay had the basics to sustain life but little more. We slept inside overseas shipping containers with three homemade bunks in each. Six men crammed each one. The container's tops were reinforced with layers of sandbags. Below the East Tower was the command center and a makeshift aid station. A few meters to the front of the East Tower stood the MWR (Morale, Welfare, and Recreation) shed. Inside, you would find a television mounted on the wall, tables and chairs, five computers, one telephone and a coffee maker. Rat-fucked care packages containing junk food and hygiene products sat on a counter. A tiny kitchen stood beside the MWR, and a mortar pit with 60 and 120mm mortars was staged in the COP's center. At the bottom of the North Tower, there was an itty-bitty prison-like gym, where legendary workouts took place.

A generator gave us electricity, but it required around-the-clock diesel fill-ups. There was no running water. When not in the bush, I stripped off my clothes and wiped myself down each day with baby wipes. We pissed inside PVC pipes, aka piss tubes that rose from the ground. Homemade stalls made with plywood were for taking a dump. Nude female pinups and pornographic mags served as bathroom décor. Everything shitted and thrown away was disposed of and burned inside the pit near the North Tower. The guards inside the North Tower breathed toxic fumes. We slept meters from the burn pit too.

COP Malekshay stood in a strategic location that made it difficult to attack by VBIED (Vehicle-Borne Improvised Explosive Devise)—aka a car bomb. The one roadway to the COP was curvy, uphill, and visible from the North and South Towers. A suspicious vehicle trying to breach the compound would be gunned down by a storm of lead. Piercing shrapnel from the claymores would get the enemy if the .50 cal machinegun didn't do the trick. There were flaws, though. While COP Malekshay stood high above the surrounding villages, it wasn't the highest point in the area. The enemy could conceal themselves inside the thick forest and lead small guerilla attacks. Their preferred method to attack the COP was by launching rockets and mortars from the surrounding hills. It was impossible to prevent rocket attacks.

"Saddle up, boys," Stan said. "And lock and load." Second Squad was heading out on our first patrol.

Squad patrols included squad members, the platoon sergeant or platoon leader, an interpreter, a medic, a forward observer (FO), and a two-man gun team from Weapons Squad. Hix and Sammy the lieutenant, joined us since we were new to the area. Sergeant Hix was the acting platoon sergeant because Manny Fresh remained in Alaska with his pregnant wife. Manny wouldn't arrive in-country for another couple of months, which pleased the squad leaders.

With a grin on my face and exhilaration in my bones, I charged my M4 rifle and walked out the makeshift gate. Most guys were excited and a bit nervous. Alpha Team was in the front, and Stu walked point. Bravo Team followed. We moved along in a wedge formation, shaped like an arrow, with Stu serving as the arrow's tip. He carried an M14 rifle firing 7.62mm rounds. Blackie walked behind and to Stu's left grasping an M203 (M4 with grenade launcher mounted). Below and to the left of Blackie, I patrolled the arrow's western edge. Dottie marched to Stu's right, carrying the M249 SAW (Squad Automatic Weapon), and Stan led the patrol from between the fireteams. Hix, Sammy, Doc Rodrigo, Trubisky the FO, the gun team, and the interpreter were dispersed throughout the formation. Bravo Team's gunfighters were flipped opposite Alpha's to ensure both sides of patrol housed a grenadier, machine gunner, and rifleman. It's not wise to have all the firepower positioned to one side.

It was a wet, dreary day, and wisps of fog hovered through the mountains. We avoided roads and patrolled the forest. The forest was thicker than it appeared from the towers. Many mature pines dotted the landscape. The ground was full of rocks—rocks for days. One clumsy step could twist an ankle or knee.

"Dude, you gotta spread out more," Stu said during a security halt. "One fucking mine, rocket, or grenade can't take out half our team."

I knelt behind a tree for cover. "Get some decent fucking cover," Stan whispered. "What the fuck is that skinny little tree gonna do?"

We headed home after a few hours roaming the countryside. I kept mental notes regarding everything from terrain to the adjustments I needed to make to my gear. No incidents nor contact with the enemy made this feel like a dry run. Stan gathered the squad, doled out critiques, and offered advice for future movements. We were wet behind the ears. But it didn't take long for the squad to operate with cohesion. When you live together 24/7, you learn and understand one another, whether you want to or not.

A routine evolved. We pulled four-hour guard shifts inside a tower with another soldier. Squad-sized foot patrols pushed out daily, putting each squad on the move once every three days. Manual labor was part of the deal too. Improving and refortifying the COP was an ongoing thing. Shovels, sandbags, and concertina wire appeared in my dreams. We did a lot of nothing also. The boredom was strong, but the "prison gym" was a godsend for fighting boredom. The stuffy little container, with rusted equipment and a life-sized Jenna Jameson poster behind the bench press, became my sanctuary as it was for many men.

Hix and Sammy were adamant about squad patrols. It got us out and seen. Platoon leadership felt activity made us less vulnerable. They believed the unit we'd relieved had been too stagnant and never engaged locals. Relationships were never formed, so attacks went on undeterred. When a local man mentioned we were the only soldiers he'd seen besides the bearded ones, it confirmed leadership's suspicions. No matter how many hands we shook, though, it couldn't solve the problem that Pakistan presented. Malekshay was located three klicks (kilometers) from the Pakistan border. Fighters could flee to an enemy sanctuary off-limits to American forces.

The drama kicked off on our seventh night. While I was lying in bed reading Maxim Magazine, Wimes appeared, looking startled. The high-strung, nineteen-year-old from Ohio stammered.

"Ggg . . . get your gear on, mm . . . man. We're bb . . . being attacked."

"What?"

"They're firing rockets," a voice yelled from outside.

Blackie and Sergeant Fink, a team leader in First Squad, tossed their gear on. I followed suit.

"You guys need to link with your squad," Fink said.

Stan slept next door, and I should've looked for him there. But he'd said earlier to meet inside the MWR if anything unexpected happened. I guess it was nerves because I chose the MWR rather than peeking inside the container steps away.

I snatched Blackie. "C'mon man, let's go to the MWR!"

We grabbed our weapons and hauled ass under the dark sky. A deafening blast erupted somewhere between the mortar pit and MWR. The ground shook, and a flash of light blazed to my left. We entered the MWR in a swirl of dirt.

"Dude, you okay?" I asked Blackie. "You see that shit?"

"Yeah, man, fucking crazy!" We stood in the MWR alone.

"Where the fuck is everyone?" I asked.

"I dunno, dude."

We were red-faced and nervous. Sweat oozed from our pores as we stood around, confused.

"Ain't Second Squad supposed to be here?" I asked.

"Man, I thought so."

Blackie and I paced in circles and fidgeted. We'd almost gotten squashed like a flea. An NCO from the prior unit appeared. He'd remained to show our platoon the ropes and was leaving the next morning.

"What are y'all doing?"

"We're trying to find our squad," I explained. "We thought we were supposed to meet here."

The sergeant laughed his ass off and said, "Y'all go back where you came from."

"Ready . . . set . . . go."

Blackie and I sprinted off again. Rips from the .50 cal thumped from the East and South Towers. Stan was the first person we met by the bunks.

"Where the fuck have you guys been?"

After we explained, Stan chuckled. "Glad you two are alright, but next time hang tight before running off like fucking cowboys."

Machinegun fire ceased, and the attack was abruptly over. Artillery had been fired from the Forward Operating Base (FOB) Boris at the suspected enemy POO (Point of Origin) site. FOB Boris supported our platoon, and most of Charlie Company was located there, seven klicks from Malekshay. The artillery fire soon stopped, and quiet returned to the starry night.

Hix found Blackie and me and laughed. "I heard you two dudes almost got smoked."

Lieutenant Sammy later dropped orders for each platoon member to receive the Combat Infantry Badge (CIB). This badge is earned purely for being shot at and taking effective enemy fire. It's every grunt's wet dream. The platoon's NCOs earned theirs in Iraq. Now the entire platoon was decorated after one week in-country. Infantrymen used to spend enlistments and careers in the army without receiving the coveted badge.

Most of us felt lame for earning our CIB through indirect fire rather than a vicious firefight.

"You'll really earn that motherfucker soon," Stu said, following the drama. "Everything comes full circle, Ronny B."

I sat alone under the stars and fired up a cig. I was not a smoker until I was in Afghanistan. I puffed and reflected. When complacency sets in, the enemy strikes. I went to bed, knowing my combat action had begun.

CHAPTER 5: EVERY DAY IS MONDAY

Every day is Monday. There's no working for the weekend, no anticipation of a hot date nor the gratification from one. There weren't thoughts about cold beer and vegging out on the couch. We lived day by day. Life was simple.

The mortar guys emptied pissers and burned feces and trash. I guess it was the lieutenant's gift to them since he never sent them on patrols. Manning their system was priority number one for them. Chores for us line guys included chopping wood and cleaning pots and pans. Everything required to keep the COP functioning, pallets of bottled water, boxes of food, and other supplies, was flown in via helicopter. The Russians scrambled like they were ready to leave as soon as they landed. Supplies often littered the HLZ as they left in a hurry.

With limited supplies, maintaining good personal hygiene was a challenge. Sweat-soaked clothes almost melted onto your skin. Your underwear and socks were constantly saturated. I was stained gray from burn pit smoke and smelled like feces and trash. Ears, ass crack, and balls brimmed with muck. I wore the same t-shirts and underwear for days because supplies were limited. Dirty laundry wasn't cleaned regularly because it was done the old-fashioned way. Clothes were dumped inside a bucket of water and scrubbed by hand with soap. Once cleaned, the clothing was hung to dry onto a makeshift clothesline.

We used archaic tools for our personal hygiene. I brushed my teeth with bottled water and shaved with razors I hoped to receive in care packages. Palletized bottled water simmering under intense sunlight made for an okay shave. I took whore baths each day with baby wipes and splashed water. Razors and baby wipes were solid gold inside care packages.

Society has this dazzling vision of infantrymen. We imagine strong men fighting for glory and performing high-risk missions heroically. Truth is, the infantryman is a grunt. We live in conditions unbearable for civilized people. The grunt lives in the "suck." The suck is a major part of a grunt's job description. Modern society produces few Americans who desire to live this way.

Each night, we gathered inside the MWR to eat. Following meals, we watched television and movies, browsed the internet, played poker, and shot the breeze. Until a field cook was dropped on the outpost, Julian was the one who cooked. He was a bright twenty-something from West Virginia with a wife and newborn baby. Straight and spiritual in his Christian faith, Julian didn't have many vices. He wasn't a party guy.

"What'd you like for dinner tonight, Bruce?" Julian asked each day.

"How 'bout filet mignon," I'd reply, dripping sarcasm.

Marty, a nineteen-year-old from Ohio, joined Julian in the kitchen. He dug a bottle of whiskey and his country music. We would sing a little old skool country and drink Wild Turkey inside the barracks. Whiskey never prevented Marty from being a hard worker, ever.

"Whatchu holdin', Phil?" I asked another comrade at the card showdown.

"I called your bet, motherfucker!"

"Aces over Jacks." I showed.

"Fuck you, bitch." Phil flipped a three of a kind I'd failed to read.

Damnit.

Phil was an older, late-twenties soldier from Michigan. He was a prankster and masterful internet troll full of one-liners, clap backs, and yo' mama jokes. Odd for a married man with three kids. But he was a good soldier who knew when to turn it up or stand down.

Our poker games were spirited. Most of the poker sharks came from Weapons and Third Squads. The regulars at the table were Phil, Smeed, Nate, and Mac. Nate and Phil were weapons guys under Smeed's tutelage. Poker was great for the boredom, but not every day was dull.

"We've gotta mission," Stan announced. "Everyone's meeting in MWR."

Sammy coordinated a mission with the company commander, and it involved most of the platoon. The squad patrols weren't enough for Sammy's resume. During our three-day mission, we'd hike to the village of Malekshay, engage the locals and conduct a shura. A shura is a meeting in Islamic societies where decisions are made. Following the shura, we planned to patrol the surrounding mountains looking for enemy positions. The guys left on the COP would perform four-hour guard shifts with four hours off. I didn't envy them. I cringed at the thought of an attack on the COP's skeleton crew, who had no medic and no leadership to support them.

After the short walk to the village, Stu and Parthia organized a security perimeter. Sammy, Hix, and the squad leaders left to plan a shura the next day. I knelt at the corner of a towering mud wall of a kalat and scanned the area. Goats, donkeys, chickens, and cattle roamed around me. I felt like a character from an ancient biblical passage because my surroundings were so primitive. Emaciated dogs limped about and sniffed dirt. Curious locals surrounded us. The local men wore what appeared to be man dresses with a sleeveless vest worn over the top. This male clothing is known as "perahan tunban." The locals had beards on their faces, sandals on their feet, and pakols, little wool hats, atop their heads. Kids cried out in English, "Mister, mister," before requesting something in their native Pashto. I carried pens and candy for them. They loved pens. American kids would throw a fit if a pen was offered as a gift.

A boy who looked twelve or thirteen approached while I puffed a Marlboro. "Mister, mister," he said as he pointed to my cigarette.

I peeked around, laughed, and handed the boy a cig. He motioned for a lighter. "Want me to smoke it for you, too?" I asked.

I reached out and lit the boy's Marlboro. He smiled, tapped his chest in appreciation, and walked off. *Clerks*, a cult-classic flick from the nineties, came to my mind; the cashier in the movie sold a pack of cigs to a child. I had mixed feelings, but it's hard to tell a kid "no." *Fuck'n A.*

"Damn, these are some fucked up people," D-Rum, a good ole redneck nineteen-year-old from South Carolina, said. "How the fuck do they live like this?"

"Yeah, we ain't in Dixie no more." We laughed.

This was one of the poorest villages in one of the most secluded societies on the planet. The people's living standards made an American trailer park, ghetto, or federal home look like Beverly Hills. The region didn't qualify for third-world status, in my opinion. Mud homes, no paved roads, no electricity, no running water, and no signs of modern technology. I couldn't imagine the area looking any different five hundred years ago.

We gathered and patrolled the surrounding mountains and ridges. This is when your grunt training is put to the test, hiking several klicks at 8,000 ft through rugged terrain. You march like a mule wearing a heavy, uncomfortable load. The most crucial test in grunting is this, the mule test, and you pass without firing a shot. The CIB is shiny, but the mule test is real. A grunt must make it to the firefight.

Nearby Rocket Ridge had earned its nickname for being a well-known enemy launch site. It possessed a bird's-eye view of COP Malekshay and FOB Boris. We hit it up. Dense forest covered the land, allowing the enemy to operate unmolested. We discovered suspected launch sites. The grid coordinates were called into the company TOC (Tactical Operations Center) on Boris. The areas weren't fortified like a bunker. There wasn't anything we could destroy that couldn't be rebuilt by evening. As we poked and prodded the Taliban land, they didn't attempt to take us on. But they had eyes on our every move. Our interpreters monitored their radio traffic and passed the info to the patrol. They saw us, but we couldn't see them. It's like chasing ghosts.

That night, we dug in high above the village of Malekshay. I shared a hole with Smitty, the fireteam's new SAW gunner, who switched roles with Dottie. Smitty grew up in Connecticut and loved modern country music, which was odd for a half-Italian kid from the northeast. He and I worked out a sleep schedule, and the platoon dialed down for the evening. I scanned the area through Night Vision Goggles (NVGs), looking for movement and ensuring no one breached our perimeter. I studied the village of Malekshay. The night was cool and steady. *C'mon Taliban man.*

On the mission's final morning, we awoke early to go back to COP Malekshay. Something wasn't right, though. A chilling sweat and nauseous feeling almost overwhelmed me. I stumbled along the mountainside, doing my best to keep up. Sweat poured and drenched my armor. Snot flowed from my nostrils. At the base of the mountain, we took a scheduled security halt outside the village of Malekshay.

"You okay, old man?" Stan asked. "I could tell something wasn't right watching you walk down the hill."

Doc Rodrigo, our young Puerto Rican medic, examined me. "Yep, you caught something in the bush, my man."

Luckily, First Platoon had driven from FOB Boris and parked near the village for the mission's duration. The first sergeant and the Charlie Company medic accompanied them. The doc's name was Doc Chauncey, and the men loved him.

"Waz up, Brucey?" Doc Chauncey always called me that.

"Ah, you know," I replied without much of a voice.

"He caught sumthin' in the bush," Doc Rodrigo chimed in.

"Well, Brucey." Doc Chauncey prepared a needle from his medical bag. "Roll up your sleeve." He then inserted the needle and administered intravenous fluids (IV).

Sammy was upset at the delay. I heard Stan tell the lieutenant, "He's fucking sick, sir. It's no bullshit!"

Doc Chauncey administered two more IVs. I needed tons of juice because we had a few more uphill klicks to hike.

The first sergeant asked, "You want him to ride back to Boris with us?"

Stan asked if I was good, I said, "I'll make it, sergeant." Off we went.

After two security halts and some brief meet-and-greets in tiny villages, we arrived home safe. I collapsed when I entered the COP. The illness became known as the Afghan Butt Flu, and it spread throughout the platoon. Symptoms included high fevers, cold sweats, nausea, vomiting, and diarrhea. Bedrest was the only remedy. The unsanitary conditions we lived under made it impossible to stop too. But I never caught it again. Those poor mortar guys had a ton of butt-piss to burn.

"You okay, colonel?" Joey C asked, after a chuckle.

"Fuck you," I replied with my weakened voice.

"The colonel" was one of my nicknames. It originated during my college years when my roommate, who served in marine reserves, called me the colonel. I called him sergeant major in response. Joey C picked it up when a young lady addressed me as the colonel inside an Alaskan restaurant. The colonel was the man of the evening, my post-midnight alter ego. Hix and some of the leadership started calling me the colonel too. My older age and smack talk had earned the nickname.

After I recovered from the Afghan Butt Flu, I resumed working out with my regular partners Shane, Ross, and Frankie. Shane and Frankie were both team leaders from Third Squad and veterans of the Iraq_War. Shane was a good-sized guy from Pennsylvania who'd wrestled his entire life. Shane's trademarks were a wad of dip and bedtime stories of banging fat chicks. Frankie was a taller slender guy of Hispanic descent. Despite being lanky, he was cock strong and understood how to use his leverage. My other workout buddy was Ross, a cool dude from Texas. Ross arrived at the unit slender. He worked out like an animal, ate supplements, and vastly expanded his slim frame.

We slung iron while others punched mitts, grappled, and shadowboxed. We tweaked workouts based off of what we read in *Muscle and Fitness* or *Maxim*, two popular reads inside guard towers.

Days crawled by, and boredom set in. Nightly conversations involved girlfriends, former girlfriends, and friends with benefits. We talked about home and what we'd do when we returned. Most of our aspirations were pipedreams, but a place like Malekshay brought out the dreamers in us. D-Rum hoped to start his own gig as a contractor building privacy fences. Wimes wanted to make a decent wage moving dirt with large excavators. I talked about starting a redneck business and maybe obtaining a VA loan to purchase my redneck biz equipment.

"Hey D-Rum, we should start our own redneck business building fences." I said as D-Rum and I smoked cigs and stared at the stars. "You do the labor, and I'll keep the books."

"Fuck off, dude. I ain't your bitch." He replied.

First Platoon convoyed to the COP each week from FOB Boris to deliver mail. On the first mail call, I drove First Platoon's lieutenant up the hill to the COP. He mentioned. "Damn Bruce, you're gonna feel like you're in heaven when you make it to Boris."

"Oh yeah, sir?"

He talked about warm showers, a chow hall providing three meals a day, laundry drop-off, a barbershop, and a Haji store with everything from cigarettes to bootleg DVDs. On Malekshay, if we ran out of cigs, we traded for them. Guys begged friends and family to send cartons of smokes inside care packages. Run out of cigs and you were "sucking dick for beer money." I once bummed a smoke from a villager, and he handed me a pack of Pines, the old Haji brand.

Knowing about the amenities that FOB Boris possessed made us resent First Platoon. To my astonishment, it'd been decided Third Platoon would remain on COP Malekshay for the entire deployment. The company would arrange for single fireteams to spend a few days at Boris in rotations.

"You gotta be shittin' me," I said to "The Duke," a team leader from First Squad.

"Yeah, that's fucked up," he replied. "I'm glad they're taking warm showers and eating ice cream there."

Care packages are necessary for some, but not so much for others. FOB Sharana and Bagram had coffee shops, Burger King, Pizza Hut, Dairy Queen, and tons of other frills. At Malekshay, care packages were our lifeline.

"Sweet," I said with excitement. "I got the Ric Flair disc I ordered!"

Digging through more mail, I discovered that Jane had sent me a portable DVD and mp3 player. *Thank you, darlin'.* The Hulk Hogan video that I sent to Frankie in secret also arrived. He was surprised and appreciative. We'd often argue over who was the greatest professional wrestler of all time. Of course, my man was Ric Flair. Frankie and his buddy Jordy screamed Hulk Hogan while shouting, "Whatchu gonna do, brother!"

My reply was, "Woooooo!" and a good Flair line,

"Limousine ridin', jet flyin', stylin' and profilin'."

"Custom made cloths, lizard shoes, and limousine a mile-long full of beautiful women just dying for a Real Man." We laughed. It was cheap entertainment that many folks take for granted.

I sorted through more boxes and found a package from my father. Thank God he sent the Marlboro Lights and baby wipes I'd requested, along with a bottle of ibuprofen. A hand-written letter was folded inside. I opened the letter, which was written on a page torn from a spiral notebook. Big RB spoke about finally receiving a one-hundred-percent disability rating from the VA. His diagnosis of pulmonary fibrosis, likely due to exposure to Agent Orange, had also been confirmed. I knew the prognosis for his condition wasn't good. He mentioned that my older sister Shelly was doing well, dating a new guy who seemed respectable. I then reached the important words. "Don't be a hero. And remember, don't be afraid to pull the trigger." It ended. "Love, Your Dad." I chuckled since the letter was written in a tone that only my father could convey.

I never received much mail. But technology surged in 2009—emails, instant messaging, and video conferences. People were fascinated to learn that they could receive instant updates from the battlefield via instant messaging and a new phenomenon, Facebook. My sister encouraged me to create a profile because it'd be easier to stay in touch. There was no shortage of techno-savvy millennials who could help me. Through Facebook, I reconnected with high school and marine buddies, albeit from a warzone. Since my father didn't have a computer, I'd message letters to my sister for her to print and give to him. I joked about being able to send and receive instant messages, yet I couldn't bathe.

We realized that correspondence from the battlefield could be passed instantaneously to the other side of the world. This led to changes in policies concerning operational security – be careful what you post and disclose.

Other people still received care packages regularly. D-Rum received *Religulous* featuring Bill Maher and planned to play it in the MWR after evening chow. It was packed. I'm not sure that many guys were fond of Maher's philosophies, but everyone appeared entertained. We laughed, provided commentary, praised and critiqued the film. Religion, and lack thereof, was debated.

As the credits rolled, a voice cried out, "Religion is a bunch of bullshit. It's good for nothing!"

"Except hope," Julian said with a quiet calm.

I'm willing to bet half the platoon and company identified as atheists or agnostics. It was a reversal of philosophies from my marine corps days just a decade ago. "No Religious Preference" was engraved into most dog tags I saw. That'd be unheard of in the 1990's military. Mine read, "Baptist Church." It'd been six years since I walked inside a church during my wedding. I never attended much anyway. But I was baptized and still representing.

Joey C said it best. "Anyone who says there are no atheists in a foxhole never served in Third Platoon."

Pondering religion inside a warzone led me to think about mortality. Death loomed everywhere. I wanted to know what happens following this life. Truth is, I wasn't one-hundred-percent sure. I wished I'd worked that out before arriving in-country. *What about the enemy?* They're solid in what they believe. The Islamic faithful are rewarded with an afterlife in heaven. A Jihadist looks forward to seventy-two virgins following their martyrdom. *If you're convinced your warrior death brings an upgraded afterlife, does that lead one to fight harder*? Who gives a fuck about death if paradise is on the other side? At best, in my mind, heaven awaited after death. But I wasn't one-hundred-percent sure.

On a clear spring morning, Parthia and I stood guard inside the North Tower. A "bird" flew in and landed at the HLZ. We joked about the last

Russian bird that crashed outside the HLZ. No one was injured, and more contractors arrived to bury the chopper where it wrecked. Suddenly, a gentleman stepped off.

Parthia peeked through binoculars. "Man, what the fuck?"

"Waz up?"

"It's fucking Manny Fresh." Manny was back, and it didn't ring music to many ears.

"Everyone into the MWR," Stan told us. "We're receiving the Op Order for tomorrow's mission."

Several of us were about to embark on our first mounted patrol. I was anxious to get outside the COP for a few days. Sammy laid it all out at the MWR. We were to conduct prearranged shuras, set up roadside checkpoints, and engage locals. Sammy wanted to hear their concerns and attempt to extract information about the enemy. A platoon of Afghan soldiers from FOB Boris would accompany us.

"Saddle up, boys," Sammy said. "Let's ride."

As Sammy's driver, I cranked the Humvee. Doc Rodrigo and Trubisky rode in back while Smitty gunned in the turret. Our convoy had four Humvees with twenty men and four Ford Rangers packed with Afghan soldiers. They carried AK 47 rifles and RPGs (Rocket-Propelled Grenades). As we exited the COP, I thought about a friend who served in the 2003 initial invasion of Iraq. He described an endless convoy stretching further than his eyes could see, a show of America's military force. Our small convoy was no such thing.

We rounded through mountains and descended into the flatlands. Third Platoon was in the lead while the ANA (Afghan National Army) element followed. I noticed Afghan drivers pulled off to the side of the wadi and stopped until our convoy passed. When we entered each village, large groups of males were gathered and awaiting our arrival. The landscape of the flats was much different than the mountains. Few trees and vegetation existed besides the occasional bush. Sammy held his meeting with village elders as they sat "Indian style" on Persian rugs and sipped chai—a warm tea that was an Afghan fav. The villagers voiced their concerns, and Sammy took notes. The lieutenant then spoke his piece, but his words fell on what appeared to be deaf ears. Pleasantries and empty promises were exchanged, and the convoy saddled up to repeat the process.

Between shuras, the platoon set up checkpoints along the wadis. Like a good occupying force, we stopped and searched every vehicle. ANA soldiers conducted searches while we supervised. Passengers were patted down while their trunks and hoods were opened and searched for devices. It appeared as if the ANA was in control. But they weren't.

Though they were intrusive, there was a reason for the checkpoints. The enemy smuggled weapons and ammunition across the vast and mountainous Afghanistan/Pakistan border. The relative safety of Pakistan made it possible for Taliban forces to resupply there and cross the rugged passes to retool their armies. When these supplies made their way across the Paktika Province, they'd pass through to the north and west and move into the Hindu Kush Mountains. At that point, supplies could be dispersed throughout every corner of the country. If we disrupted the initial flow, we'd better ensure our safety and the safety of countless others. However, it was impossible to accomplish with twenty men and our inept ANA comrades. Surrounding spotters also radioed the location of our checkpoints to Taliban smugglers so they could avoid them.

As evening drew near, we hunkered down into a security perimeter and called it a night. Our ANA counterparts drove to the bazaar and purchased fresh fruits, vegetables, and goatmeat. Smoke fluttered from a giant black pot that they used to cook each evening. It brought back memories of camping in the Appalachians with Wildfire. Toking a Pines, I relaxed outside the Humvee.

An ANA soldier approached with a teacup in hand. "Chai?"

"Sure."

"You gonna drink that shit?" Smitty asked.

"Hell, I'll give it a try."

Ah, the warm chai soothed my throat. The taste was refreshing and surprised me. Coffee-warm, Afghan chai isn't the cold, sweet, iced tea of the South. But it was smooth and meshed with my cig.

"Hey gents, come on over, and grab a bite to eat," the lieutenant announced.

Smitty, Trubisky, and I followed Sammy to the feast. We sat side-by-side, Indian style, amongst ANA soldiers on bright Persian rugs. They brought huge tin trays piled with goatmeat, rice, and vegetable stew. The other ANA dudes brought plates of circular-shaped homemade bread. I felt like I was being served inside a restaurant that, despite serving delicious food, fails the health inspection every year. The ANA soldiers were courteous hosts and insisted that we eat until stuffed.

"Any utensils?" I asked Stan, who sat across from me.

"You don't get 'em here."

The Afghan to my side cuffed his sleeve and used his hand to scoop a chunk of stew from the plate to our front. He then stuffed it into his mouth.

"Guess you don't get your own tray either," I said and laughed.

Fuck it. I used my own filthy hand to swoop in on the goat stew from the same tray the Afghan soldier ate from. *Fuckin' A; damn good.*

Then it dawned on me; I hadn't bathed in over a month. I'd be willing to bet the Afghans to my side hadn't bathed in forever, either. It was this or a Meal Ready to Eat (MRE), and those were getting old. So, I had no qualms about eating with my dirty hands as the greasy Afghan next to me ate from the same tray. Other Afghan soldiers tore pieces from the large circular bread and offered one to me.

"Yessir." The homemade bread was tasty, as good as any bread I'd eaten.

"You know they cook that over goat shit, don't you?" Stan asked.

"No fucking way?" I replied.

"Yep, they sure do."

Stan went on to explain the process of how Iraqis cooked the same type of bread. I didn't give a fuck. When you've been living the suck this long, a petty thing like eating shit bread doesn't cause you to bat an eye. But I figured Stan was pulling my leg.

Another Afghan handed me a slice of watermelon to finish things off. Home-grown, it appeared smaller than Florida watermelons.

"More?" the Afghan asked.

"Naw dude, I'm good," I said and waved him off.

I saw femininity in these Afghan soldiers, which struck me as odd. Their AK 47s had bright-colored flowers and roses drawn on the magazines. Their fingernails were painted orange, blue, or pink. Most flaunted their femininity. This was a far cry from the rough, fighting men, I had envisioned, who defeated anyone who dared to invade. I sensed something fishy about these guys.

"Damn, are these guys homo, or sumthin'?" Rusty asked.

"Guess so," I said, and we laughed.

Rusty was an easygoing nineteen-year-old who moved around a lot as a youngster, splitting most of his time between Tennessee and Texas. I took him under my wing. Some people said he was crazy, perhaps he was, but he was intelligent and thought critically. We watched Charles Manson clips on YouTube. He also liked the old skool country music I played; honky tonkin' Gary Stewart was his fav. I was happy to see a young man avoiding the modern pop country garbage.

On the last day of the mission, we packed our gear to RTB (Return to Base). The ANA soldiers split and drove towards Boris. We were headed for the mountains and COP Malekshay.

Staff Sergeant Rubio greeted me before we left. "Was sup Ronny B, baby? Stylin' and profilin', woooo!"

Rubio heard the guys refer to me as the colonel and ole Ronny B. He'd also heard folks greet me with, "Waz up, baby?"

"Sergeant Rubio, what's goin' on, baby?"

Rubio and I chatted for a few minutes. He went on about how lame this mission was and that he wanted to kick in doors and shoot some fucking Hajis in the face. He meant it too.

Rubio was a squad leader for Third Squad and a hardened combat veteran with several tours in Iraq under his belt. He'd been shot several times in the leg during his first tour. Rubio spent months inside a hospital and could've received a medical retirement at one hundred percent. However, Rubio was a war junkie. A few bullets wouldn't keep him from continuing to fight for Uncle Sam. Rubio was that one percent I'd heard Big Ronny B speak of in Vietnam, the ones who loved war. He had enough time and service to be a sergeant first class leading his own platoon. Yet he wasn't the textbook garrison soldier. He was the "break glass in case of war," type.

"Round it up, fellas," Sammy said. "We're heading home."

On the drive back, we avoided wadis and embarked on a four-wheeling adventure. The Humvee bounced and climbed. Soon we reached the base of the final mountain and the curvy dirt road that went up towards COP Malekshay. All of a sudden, the ground beneath us shook as a blast erupted. Dirt and small chunks of rock fell from above, peppering the windshield.

"Keep moving; keep moving," Sammy shouted. "Scan your sectors. Smitty, what's going on?"

"The vehicle behind us has been hit, sir!"

We drove further up the hill before rushing back on foot to check on our comrades. The vehicle's hood was blown off, and the front end was destroyed. The tires were flattened, the bulletproof windshield cracked, and burn marks peppered the front side. *Thank God no one is hurt.* We cordoned off the area and pushed a small patrol into the mountainside. The perpetrator was nowhere to be found. As we drove back to the COP, I worried another Improvised Explosive Device (IED) might be in our path. There wasn't.

Most IEDs are detonated by one of two methods: a pressure plate or command. A pressure plated IED reacts like any mine. Explosions occur when weight is applied. Command-detonated IEDs are triggered by cell phones or other electronic devices. We didn't know which type of IED struck our convoy.

"Man, I was worried about you." Joey C said, following the mission. "I was in the North Tower and watched the whole thing go down."

I lit a Pines as I listened.

"Thank God my lil' buddy is alright." He concluded.

"I was worried about a secondary explosion." Stu nestled into his seat with his legs crossed. "I could've sworn it was coming after we dismounted the Humvee."

"That would've been rough." I kicked back and relaxed too.

"Lemme get a cigarette," Stu said. "I sure could use one."

Health nut Stu was a nonsmoker. But all bets were off. It was disheartening knowing the platoon's first mounted patrol ended with an IED blast. But we were all in one piece. *Think I'll hit the MWR and go online to see what's happening in the world.*

CHAPTER 6: MAN DOWN, DRIVE ON

Winter fizzled, and the remaining snowcaps melted from the surrounding mountains. Two to three dudes went home each month for a two-week R&R (Rest and Relaxation). The platoon size shrunk, and we went down another man gone AWOL (Absent Without Leave). It was a fellow member of Second Squad named Phife.

"Yeah, it sucks," Stan said when he gathered the squad to pass the word. "But we're gonna have to drive on without him. If he wants to be a bitch, we don't need him. Fuck him; he's dead to us now."

"Fuck that pussy," Blackie blurted out.

"Deserter, the ultimate blue falcon (buddy fucker)," mumbled from somewhere.

Phife was on the first scheduled R&R rotation and tapped out somewhere between Afghanistan and Maryland. *What a bitch.* A whopping two months was all he could handle before he broke. *I'd have a tough time looking at myself in the mirror.*

Life went on, nonetheless. Bonds strengthened. We continued with playful ribbing, arguments over which band or sports team was best, talking about female conquests and all that. Stu referred to the eighties metal I listened to as "butt rock." He tried to convince me that Killswitch's version of *Holy Diver* was better than Ronnie James Dio's original.

"Cracka pleez," I said and made my "whatever" face. "Who the fuck is 'kill bitch?'"

A decade-plus age gap existed between many platoon members and me. I was Generation X, and they were Gen Y. My idea of metal was Motorhead, Danzig, Priest; theirs was something like Five Finger Death Punch. My idea of country was Hank Jr and III, Waylon, Conway, George, Emmylou, and Crystal Gayle. They were into Paisley, Chesney, Luke, and Aldean. Some liked Hank III since he was recent. I could dig some of their rock and rap but despised modern country music.

Over the years, I'd racked up over a hundred concert appearances – everything from local bands to headliners. I'd seen Danzig five times, Hank Jr five times, Dio three times, and Motorhead a couple times. Nashville Pussy was one of my favs; I'd seen them a dozen times since they were Atlanta-based. Blackie dug NP, too. Guys loved hearing about my concert adventures because they hadn't entered that phase of their lives yet.

Sometimes I'd ridicule the fellas and their generation like salty older people tend to do. Throwing shade at one another was essential biz on deployment. Joey C and his buddy Harlen never missed out on the action. Harlen was the youngest platoon member, eighteen, and a touch shy. He was Filipino and white, grew up in Utah, and loved international soccer. He dug rasslin' too.

"You old bald-headed fart," Harlen said. "You gonna need a walker climbing that mountain tomorrow?"

"Bitch, if I had Asian hair like you, I'd rather not have it," I replied. "Daddy's legs work fine."

You have to take it as well as give it. Thin skin is a sign of weakness here.

"Daddy was fingerin' his girlfriend in high school while you two were infants sittin' in a baby's seat in your mama's Honda." I leaned back on the bench, puffed my cig, and smirked.

"Yeah, yeah, old man, at least our mama's car was nice, unlike that raggedy shit your mama drove." Joey C stood and flexed Hulk Hogan style. "Whatchu think 'bout that, bitch."

"My mama didn't drive," I said and laughed.

I rambled about being a toddler during the late seventies – sitting in the car between my smoking parents with my mother's arm across my body as a seatbelt.

"Fuck you, old man. Go drink some Ensure to 'ensure' you make it through tomorrow's hike," Joey C said. "Hopefully, you won't shit your Depends."

Good one, Joey C.

The ribbing filled the time between the action.

"Aw hell, here it goes again."

Stu and I sat inside North Tower when we heard the shelling start. I scanned the sector to my front and noticed rounds exploding on a hillside. The enemy was "walking" rounds onto our location; they adjusted initial off-target fire until it moved closer to the target, COP. Our only hope to stop the barrage hinged on FOB Boris spinning up a counter-fire artillery mission. Sometimes they reacted quickly but most times, they were too late to be effective. Sometimes a fire mission requested was flat out denied. We grabbed our balls and weathered the storm; nothing we could do.

Denied fire missions were infuriating. Are commanders worried more about collateral damage than the lives of troops? Is a Rear Echelon Motherfucker (REMF), sitting inside a cozy office drinking Green Bean Coffee and stuffing his face with Burger King, afraid to drop the hammer? *Why don't they let our mortar guys fire their system from the COP?* It was the one available weapon capable of reaching an enemy several klicks away. We needed permission to fire it from someone higher than Sammy. Our approved crew-served weapons were out of range and ineffective versus a mortar barrage.

The shelling continued. Parthia and I were chitchatting inside the East Tower during one particular barrage. The afternoon sun was beginning to set when it began.

"Here we go, Ronny B."

Nothing could be done but look for the POO and point of impact (POI). Once we figured that out, we'd radio the estimated coordinates to the SOG (Sergeant of Guard).

Parthia and I heard Brandon over the radio. "SOG, SOG, this is North Tower."

"This is SOG; go ahead, North Tower."

"I saw three men running from a house, break. They ran from one house to another, break. Then ran into a field."

"Copy that North Tower."

Parthia and I stared at one another. "Fuckin' A, Ronny B, Second Squad is going out."

"You damn right we are, Sergeant Parthia."

Stan rushed inside East Tower with two guard replacements. "Get your shit, gear up, and meet in the MWR. Second Squad is goin' on a walk."

Sammy explained a brief op order. He and Hix would accompany the squad.

"This oughta be fun," I whispered to Joey C since this was our first time walking through hostile territory utilizing night vision goggles (NVGs).

Darkness blanketed the sky but the greenness from the NVGs illuminated the surroundings a bit. But the lack of moonlight made it hard to see well. We lacked depth perception and walking through unforgiving terrain is a hell of a way to break in NVGs. I busted my ass a couple times, and my ankles twisted and almost sprained. I couldn't make out the loose rocks that covered the ground. Clanking gear and occasional falls could be heard.

We maneuvered through a wide-open grassy field once we descended into the flatlands. In a bizarre way, the field of darkness was also stunningly beautiful. I savored the moment. My mind wandered to films and documentaries of Vietnam where soldiers patrolled through tall grass and rice paddies. Men carried weapons at the ready, mindful of booby traps or the VC prepared to ambush. Some 'Nam grunts carried the M60 machinegun across their backs. Known as "the pig," their hands dangled over its top. A thing of beauty it was, in my mind. I felt connected to my Vietnam combat elders.

We halted outside the neighborhood of our target home.

"I need you and Blackie to follow the lieutenant," Stan whispered. Blackie and I stood and moved towards Sammy in silence.

An interpreter joined us. We pushed through bushes, stepped across ankle-high mud, scaled a chest-high wall, and entered the neighborhood. The four of us knelt and huddled.

"That's the house," Sammy said as he pointed. "Stay close while I knock on the door."

With weapons at low ready, we approached the home. Sammy raised his fist to knock. My rifle rose to high ready, barrel pointed at the door. The green laser from the PEQ-2 mounted on my M4 put a green line on the door's center. My safety was off. When the door opened, and a normal-looking local man appeared, I lowered my rifle to look less threatening.

Sammy and the man jibber-jabbered through the interpreter as Blackie and I stood guard. I peeked inside the home and noticed a family huddled together, concerned by our presence. The kalat was candlelit, which further supported my notion of Afghanistan as a less than third-world country. The man agreed to test to see if gunpowder residue was on his hands. There wasn't any, so off we went. If bad guys were in the area, they had vanished.

Stan wasn't a happy camper when we returned home. "God damnit, what kind of shit was that? I heard banging and all kinds of shit. Stu, Parthia, check your guys' shit. Make sure it's fucking secured and tied down, fucking silenced!"

"Roger that, sergeant," Stu and Parthia replied.

"Fucking embarrassing, man."

Our evening poker game was interrupted when the platoon was spun up because a of a "gist." A gist, a type of intelligence, heard by an interpreter over local radio traffic, hinted that the enemy planned to attack the COP. When one came in, we knew the deal. The guys in towers hunkered behind their weapons. Some dudes manned the Mark 19s mounted on the parked Humvees, and the rest of us took positions along the wall.

Stan hunkered to my left. "C'mon, you motherfuckers," he screamed into the darkness. "I'll send you to your fucking Allah!"

Goddamn right, Iron Guts.

It was a false alarm.

Later, Stan summoned the squad.

"We've got another mission. I think you're gonna like this. We're gonna spend a night at FOB Boris!"

Boris? Fuck yeah!

"Here's what's goin' down," Stan explained. "A platoon from Boris will pick us up. They'll be accompanied by Afghans who'll stay and help the skeleton crew pull guard. The platoon will drive us to Boris. When we get there, they'll give us a quick class on the MRAP, and we'll roll out."

The MRAP (Mine-Resistant Ambush Protected) vehicle is massive and armored. A V-shaped underside allows it to deflect IEDs like a boss. The vehicle has a weakness in the side armor, though, which makes it vulnerable to RPG blasts. RPG cages, which caused RPGs to detonate outside the armor rather than directly onto it, addressed this flaw. Ours didn't have cages, though. It was a nice vehicle to ride though because the MRAP carried more troops and had increased IED protection compared to a Humvee.

We arrived on FOB Boris and had a short time to eat hot chow and chat with guys from First Platoon. I milled around with Joey C and noticed a basketball court and small gym. The laundry drop-off, barbershop, and Haji store were all there. Inside the Haji shop, I saw cigarettes, dip, bootleg DVDs, fake jewelry, and counterfeit Rolex watches for sale. We continued touring the FOB and saw dudes walking around in t-shirts, shorts, and tennis shoes with rifles strapped across their backs. The DFAC had two rooms with a television mounted in each. Open 24/7, soldiers could stop by anytime to grab an ice cream or near beer.

"Not bad, huh, Joey C?"

"I could get used to this, Ronny B."

We headed to the Afghan National Army (ANA) side of FOB to receive a brief since they were tagging along on the mission. They had their own area and weren't allowed on the main side unless escorted by a commander.

A model constructed with rocks and sticks was laid out on the ground of the ANA compound. "What the fuck is this?" I whispered to Joey C.

Two US Soldiers were speaking to the Afghans through an interpreter.

"These motherfuckers can't read maps," I said to Joey C.

"Fuckin' A, man," Joey C replied. "We gotta explain this shit by drawing it in dirt like sandlot football?"

We had a three-day mission with the ANA in the MRAPs. It was a Santa Claus mission, an attempt to offer the locals gifts and goodwill. An ANA soldier drove what could best be described as a U-Haul truck packed with soccer balls, toys, food, water, etc. There were no ambushes or attacks, but I still noticed something strange.

As we drove from village to village, we'd stop, dismount, and patrol by foot. The children approached us because they were curious about foreigners cloaked in armor and large weaponry. When I looked at the preteen and teenage boys, I noticed their eyelashes were enhanced with mascara. Their fingernails were also painted pink, and the boys acted in a feminine manner. I recalled our first mounted mission when I saw painted flowers on magazines of AK 47s. *Something ain't right.*

We completed the mission and headed to Boris for warm showers. It'd been seventy-seven days since I'd bathed. An enthusiastic group of us, wearing shower shoes for the first time in Afghanistan, filled the stalls with laughter. For a short time, I cherished the feeling of lukewarm water. Dirt and grime washed down the drain.

The following day, the plan was to drive the MRAPs to COP Malekshay. Later, a platoon from Boris would fly out and drive vehicles back to the FOB. I took my place in the back of the MRAP and chilled. Yet relaxation was short-lived. Less than a klick outside Boris, an explosion filled the air with shards of razor-sharp metal.

I saw a smoking MRAP through the windshield. "Holy shit, that's Stan's vehicle!"

The Second Squad dudes had been hit. Parthia was driving, Brandon gunned, and a few guys rolled in the back like I'd been doing. The go-ahead was given to dismount, and we rushed to the disabled MRAP. As we approached, Parthia and Stan exited.

"We're good," Stan said.

Thank God.

The MRAP took the blast like a champ. A monstrous armored tow truck drove out to recover the MRAP since it'd been damaged too much to drive. The good news? We were going back to Boris to spend another night.

We showered, ate hot chow, and enjoyed an ice cream dessert. I bought

DVDs to take to Malekshay. Firing up a cig, I sat on a bench under the chill starry night. D-Rum, Wimes, Harlen, Blackie, and Joey C surrounded me as we enjoyed another evening.

"Weren't you in Okinawa, Ronny B?" D-Rum asked.

"Yep."

"How was that?" Harlen asked.

"Damn good times."

"What'd y'all do?" D-Rum asked and puffed on a fresh cig.

"Well, we trained and partied like a motherfucker."

"Oh yeah?" Blackie asked.

"Yessir," I said. "There wasn't a war goin' on. My marine corps days were a Far East pussy fest in Okinawa and Korea. Then good times on the beaches of Southern California."

"Say what?" D-Rum asked.

"Yep, in Okinawa, you could buy a whore known as a 'skivvy mama' for ten bucks. There was a crazy-ass show called the 'Banana Show' where some gal performed on stage smoking cigs with her vagina. She even stuck a banana up there then spit it out into some drunken jarhead's mouth. Dude swallowed it!"

"No fucking way!" The boys laughed.

"Lucky motherfucker," Harlen said.

"Y'all will get yours when we get back to Alaska."

"What's the craziest time you had in Okinawa?" D-Rum asked.

"Hmm . . ." I thought. "There's one in particular that stands out." I started the story.

"Daddy was a fresh-faced eighteen-year-old who'd only been in Okinawa for a couple weeks. One night ole SilverDawg, T-Spoon, Brian B, and yours truly decided to head to a bar called The Jet, near Kadena Airbase. We young jarheads beat our chests, screamed our war cries, and pounded the hard stuff.

What's crazy is this Japanese band belted out the American rock and metal tunes of the time. They covered Danzig's *Mother*, and it sent us drunken jarheads over the top."

"Daddy then made his way to the latrine to relieve himself. A blonde American chick grabbed me by the crotch as I made my way through the crowd. *Damn honey.* She wasn't trashy as you might think; she was a smoking hot sweet young thang. I offered her greeting but, she kissed me instead. What's a man to do? You go with the flow and enjoy the ride. We made a little small talk, and I found out that her daddy was a sergeant major. She was visiting on her college break and flying home to California the next morning. Hot to trot, she asked if I'd come back to her father's house. I told my boys—'see y'all later.'

"The next thing I remember was waking in her room hungover. *Oh shit, it's 11:30am!* She was supposed to wake me at 7:00am before flying home, but I was passed out cold. I heard the voices talking in the hallway, fuck. I knew I had to get out, so I dressed without making a sound. Quiet as a mouse, I opened the second-floor window, popped out the screen, hung from the ledge, and dropped to the ground. After landing, I sprinted out of the strange neighborhood. A cabbie passed, and I flagged him down."

"He spoke no English, so I asked, "Kinville . . . Kin?"

"Ah, Kin." He drove me back to base.

"Are you shittin' me?" Wimes asked as everyone erupted with laughter.

"Fuck no," I said. "I wouldn't bullshit you, my cracka."

"Did you fuck her?" Blackie asked.

"Don't remember, never even got her name!"

We got a much-needed laugh after yet another IED attack.

CHAPTER 7: BYE-BYE MALEKSHAY, F**K YOU BERGDAHL

Back to COP Malekshay we went, and the suck resumed. One thing for sure was Malekshay made us tough as fuck. Isolation forced us to be a team. Malekshay wouldn't stand forever, though. Rumors insinuated that Malekshay would be torn down, and the area abandoned. The platoon would move to FOB Boris.

The order was passed down without warning. Frankie and I were on guard duty, arguing over the Dodgers and Braves. Near the crack of dawn, Hix entered the South Tower and broke the news. Thrilling as it was, there was tons of work to do. Frankie and I started moving boxes and clearing the storage container below the tower of supplies. The move began while most people slept. A rumble from a mortar blast awakened the platoon, but Frankie and I kept working as if nothing had happened.

Manny Fresh arrived on scene in full gear. "Woot ze fook? Gid ze fook oonder soom coover in git yoor gear on!"

"Roger that, sergeant."

The teardown of Malekshay kicked off with sounds of an explosion and Manny's screams. The project went on for weeks. We were ordered to destroy the place until nothing remained except a bare dirt mound, as if Malekshay never stood. The work was backbreaking manual labor with hammers and knives. Every wooden building was ripped down, and each sandbag was sliced and emptied. The only assistance came from local truck drivers and heavy equipment operators. We loaded the wood into large trucks to be hauled off. Local heavy equipment operators used massive excavators to destroy Hesco walls filled with dirt. I likened our labor to working a rigorous blue-collar job, but it was around-the-clock, and there were no showers before bed.

As the COP came down, nothing shaded the blistering June sun. During breaks, we laid down underneath vehicles and napped. We also discussed the over-the-top "jingle trucks" locals drove. They were painted in bright, loud colors with tassels and chains hanging everywhere. Creepy music emanated from them as they cruised the wadi. It reminded me of the ice cream man riding through the hood when I was a child. The people were gaudy as fuck like Liberace, but a poor man's Liberace.

The locals helping us tear down Malekshay camped in their own area. We noticed one dude didn't have a role in the teardown; he drove no truck nor operated equipment.

While taking shade under a vehicle, Blackie asked. "Hey Bruce, you notice that one Haji doesn't do shit?"

"The feminine one?"

"Yep, I bet he's their 'boy,' if you know what I mean."

"He's gotta job, alright," the Duke chimed in while kicking back in the dirt with a cap shielding his eyes.

"He's a fuck boy," someone blurted from under the vehicle.

One afternoon, some drama kicked up at the Haji camp. The little fuck boy was throwing hands with another dude. Shouts of, "Kick his ass," "Knee him in the balls," and "Don't let him punk you like that" rang out.

They slapped one another and grappled, using their man dresses like martial arts "gis." We egged them on. "Must've been some love triangle shit," I said after the drama died.

We continued tearing down Maleskhay. The containers were airlifted to FOB Boris once the bunks inside were destroyed and tossed out. Our sleeping arrangements shifted to lying on cots or dirt. We constructed what appeared to be a gypsy camp for our last nights on the demolished COP. Large tarps stretched overhead and protected us from sun and rain but not enemy fire. We'd be whisked away to FOB Boris soon, anyway. There was a light at the end of the tunnel. But that light flamed out.

Stan approached as I tossed wood into a jingle truck. "Stop what you're doing and gather the squad."

I found Stu and passed the word. We questioned aloud what could be so important that we needed to meet asap.

"I got some shitty news," Stan said in a serious tone. "Some kid from Blackfoot Company 501st went missing." He paused a bit and elaborated. "All operations in the brigade will cease until he is found, including this teardown."

Huh?

"I know it sucks, and there isn't much we can do," Stan continued. "So, we'll sit on this dirt until we receive further word. All our assets are focused on finding this guy."

"Hope they find this dude fast," Stu said. "Any word how he went missing, sergeant?"

"Word is he snuck off base in the middle of the night."

"You gotta be shittin' me, sergeant," I said.

"Nope, not at all, man. He said, 'fuck it' and 'peace out.' And we'll eat a shit sandwich until he's found."

"Goddamnit, man," Blackie said. "At least Phife pulled his pussy-ass shit on R&R."

"Oh, one more thing," Stan said before the huddle broke. "Don't say a fucking thing about this to anyone. Not your wife, girlfriend, mommy, daddy, or no one. This shit will be all over the news, and people will be asking about it. Don't say a goddamn thing!"

"Roger that, sergeant."

Bowe Bergdahl walked off the nearby OP (Observation Post) Mest on the last day of June 2009 while we were balls deep in the teardown. He gathered water, a knife, a compass, and a diary and stepped off into the hills. Soon after, Bowe was captured by the Taliban and held captive in the Haqqani Network. His captivity sent ripples throughout the military. The American media rushed to weigh in, spinning unflattering narratives based off misinformation. Early news reports indicated Bergdahl was left behind on a foot patrol. I knew that was bullshit from the get-go but couldn't share it. Stan once mentioned—the worst thing NCOs can be accused of is leaving a man behind. That's what the media said – he was fucked by his own platoon. It wasn't the case, though. Bergdahl's status was never classified as Prisoner of War (POW). Instead, he was considered DUSTWUN (Duty Status Whereabouts Unknown).

Following Bergdahl's disappearance, the platoon divided into two groups that rotated every eight hours. One group would patrol and dig into a position in the hills. The other half rested and guarded the gypsy camp. The rotations would continue until the missing soldier was found.

"Fuck that DUSTWUN kid," was heard a lot as we labored.

"We should be sippin' Cokes and eatin' pizza at Boris now, Joey C." We laughed and toked cigs as we choked on this shit sandwich.

After two weeks, word came from the brigade that the teardown could resume. Bergdahl was in enemy hands. *Fuckin' A.*

As work continued, there were structural changes to the platoon. Second Squad was forced to give up a team leader for a detail at a major FOB. Stan chose Parthia because he'd been "stop-lossed." It was fair for him to get a relaxed duty in the rear since he wasn't supposed to be here anyway. The new sergeant who replaced Parthia had checked in to the squad a month before deployment. Ole Iron Guts treated him like a private before dishing him off on another unit. T-Love was his nickname, and we'd reunite at Boris following the teardown. Sammy was also leaving for another assignment. The brigade commander had chosen him to be his right-hand man, aka the "colonel's bitch." It pulled him from the battlefield and sent him to an office. I knew the lieutenant wouldn't be happy handing out coins on behalf of the commander. Sammy didn't want to just lead; he wanted to make a difference.

We imagined a baby-faced lieutenant as new as a private straight out of basic would appear. A bird dropped off Lieutenant Troutman instead. He was a thirty-something prior-enlisted guy who'd had deployments under his belt. We referred to prior enlisted officers as "mustangs" in the corps. Mustangs are well-respected because they've marched in enlisted men's boots. The new lieutenant was a lot of things that the young officers weren't. For one, he smoked at least a pack a day and never hid it. Unlike the average lieutenant, he was combat tested and had little crow's feet and a touch of gray hair. Despite not looking the part, Troutman was a professional. He prioritized troops over career, the mustang inside. Before trips "outside the wire" (beyond the base), he gathered the members of the platoon and lead a prayer.

There was always something happening other than just the teardown. An ANA Ford Ranger drove over an IED on a trip to Malekshay. All seven of the vehicle's occupants were killed. First Squad was spun up to lead a patrol to help the ANA recover the vehicle and bodies. Staff Sergeant Donald, a laidback cat and workaholic type from Kansas City, led the squad.

I was leaning on a Humvee smoking a cig when Stan approached. "I'm gonna need you to help them out, Bruce."

"Roger that, sergeant,"

I was gearing up when Stan found me again. "Be careful out there, man."

"I'll be good," I said, smiled, and gave him a fist bump.

But Stan's look and words made me more vigilant. The ANA policed the scene and hauled off the bodies before our arrival. As a result of the deaths, a squad pushed out each day to provide overwatch for convoys. We set up high atop roads with the best available sightlines and scanned the area for suspicious people. Perhaps we'd catch someone digging. We were a small group searching for a needle in a haystack through nasty terrain and too many blind spots.

The action heated up when an IED popped off and struck an MRAP from our platoon. The damage was minimal, but we didn't have eyes on the situation since the explosion occurred inside a blind spot. Stu spotted what appeared to be a man fleeing the area, and the chase was on. We sprinted full bore attempting to keep the wedge formation intact as we ran. Downward through the woods, we double-timed. We emerged from the forest and raced a hundred meters down a dirt road before settling back into the trees.

Stan approached Joey C and me and whispered. "Follow me."

Slow and tactical, with weapons at high ready, we moved. Stan gave the signal, and we charged the hill. Two kids flashed into view. Stan's rifle dropped low.

"Hey, I know you two," Stan said and smiled.

He offered a handshake to the two young boys he'd encountered on an earlier mission. With no interpreter at our disposal, Stan motioned them to go away.

"Y'all get the fuck outta here."

Finally, COP Malekshay was no more. We had one final night to spend on the mound of dirt before driving to Boris in the am. Stan believed an attack was imminent. We placed MRAPs and Humvees into a perimeter and hunkered in for a fight. I nestled into a prone position in the dirt beside Joey C, nervous but ready. We heard the whistle of an incoming rocket as it flew over our heads and exploded in the woods to the south. The .50 cals and Mark 19s opened up as Brandon and Stu laid scunion (destruction) on those motherfuckers. Lieutenant Troutman radioed FOB Boris, and a thunderous volley of artillery rained onto Rocket Ridge. Our vulnerability must've guaranteed approval. Minutes later, all was quiet. I spent the evening lying prone with my eyes glued on my night vision, awaiting an enemy rush that never came.

The following morning, we saddled up for the convoy to FOB Boris. I didn't envy Stu, who sat shotgun inside a Humvee pulling a trailer full of mortar rounds. The mortar rounds included highly flammable "Willy Pete" (White Phosphorous).

"Good luck in that rolling bomb, cracka."

"Fuck you, Ronny B. Daddy'll be fine."

CHAPTER 8: A NEW BEGINNING

The days of poor hygiene and shitty food were over, or so we thought. Our new home was pleasant. The dining facility, MWR, barbershop, laundry drop-off, and Haji shop were welcomed amenities. One big difference between COP and FOB life was not walking around kitted in armor most of the time.

Stu, Blackie, Smitty, and I occupied a room with plenty of space. One day, we strolled to the chow hall. Midway to our destination Stu belted out. "What's up, T-Love?"

Our new squad mate and Bravo Team Leader had passed in the opposite direction. T-Love was twenty-three, reserved, and had a boyish look. Acne covered his face. His personality reminded me of Steve Carell's character in *The 40-Year-Old Virgin.*

T-Love wasn't the only addition to Third Platoon. Paris, a laidback African American dude from Memphis, joined us. He was one of few Charlie Company cherries to earn an Expert Infantry Badge.

When I realized Paris was from Memphis, I asked about the local musicians. "How 'bout that 8Ball and MJG?"

He chuckled and said. "Yeah, good shit. They from Memphis."

Paris was transferred to our platoon because of his issues with Second. But he was solid, and we were glad to have him. Paris also aspired to be a "lifer," someone who spends their entire career in the military.

I'd sometimes ask, "Why you wanna deploy again?"

"Cuz it's an addiction."

I almost understood because FOB life was pleasant. We added a small television and used Xbox to our room. My video game playing days were over, but other guys passed the time playing *Tiger Woods Golf* and *Call of Duty*. The FOB gym was an old rundown building, but it beat the prison gym at Malekshay by a mile. Dudes shot hoops on a basketball court while others jogged the FOB's perimeter. The biggest luxury was dropping off laundry to be done by locals; no more handwashing clothes inside a bucket. We even got free haircuts at the barbershop. Tip the Haji barber a couple US Dollars and he was all good.

When they were not playing video games or working out, guys could find other things to do. Illegal narcotics were plentiful on Boris. Afghanistan was the largest cultivator of opium and heroin, producing ninety percent of the world's supply. In 2009, the nation produced a stunning 6,900 metric tons of heroin. We were standing in the illicit drug capital of the world, and it was impossible to prevent narcotics from seeping into FOB. Dog Platoon dudes pointed out which interpreter, employee, and ASG or ANA soldier slang dope. Guys strolled to the ANA side to obtain drugs. Third Platoon's isolation atop COP Malekshay had sealed off our interactions with the dark side of hashish, pills, and smack, but Boris offered no such protection.

Our focus at Boris was overseeing an area of operation (AO) that included the Bermel District, which sat inside the Bermel Valley. It bordered one of Pakistan's federally administered tribal areas known as North Waziristan. The AO stretched from Margah to the north (not to be confused with Marjah), Malekshay to the east, a mountain range to the west, and an area known as Shkin to the south. Firebase Lilley, the exclusive home to US Special Forces, was in Shkin. Some called it the Alamo. Shkin was the front of the frontline in the war versus Taliban and Al-Qaeda fighters. Army Colonel Rodney Davis called it. "The evilest place in Afghanistan."

We were one of three infantry platoons charged with securing the AO. First Platoon was moved to another FOB and no longer worked the area. Second was stationed on a COP in the northernmost region of the AO. My platoon operated out of Boris alongside "Dog Platoon." They came from Dog Company and aided Charlie Company.

A rotation went as follows: one platoon served as the Quick Reaction Force and protected the FOB. The other led missions "outside the wire." When we were on force protection, we manned two posts and provided an NCO to serve as SOG. One of the posts was a tower located at the entrance. The other was a shack separating the American from the ANA side of FOB. When we were on guard, we were accompanied by two ASG Afghans armed with AK 47s.

Soon a mission came down the pike, and the platoon was summoned to a conference room to get the scoop.

"Lace your hiking boots, gentleman," Lieutenant Troutman said. "We're gonna walk up Spaghetti Hill."

Spaghetti Hill was a large bump in middle of flatlands with ridges that ran perpendicular to surrounding mountains to the east and west. It was a convenient enemy launch site and just three clicks from Boris. Our mission was to hike to the top, and then spend a couple days patrolling and looking for signs of enemy activity. Second Squad would lead the patrol. Stan ordered Bravo to the front, placing ole T-Love on point. It was a baptism by fire because the point man is crucial. Contrary to popular belief that new privates are thrown on point, it's really a job for an experienced sergeant. Walking point is what team leaders do. T-Love was being tested right off the rip. I suspected Stan hoped that T-Love would shit the bed in front of the entire platoon.

We stepped off in the am. I giggled over T-Love's knee and elbow pads, and bloused boots. He looked like a cherry. After a few klicks, we hit the base of Spaghetti Hill and ascended. The view from the peak was breathtaking, the best view of the valley. There were no signs of anyone living on the hill. At night, we set up a perimeter and dug in. Brandon and I were grouped together.

While Brandon and I discussed a sleep schedule, Stan summoned the squad for a night patrol. "We gotta secure this hill so let's walk."

"Roger that, sergeant."

We hiked a couple klicks into the backside of the hill, which was not visible from the FOB. The harsh terrain was brutal to navigate. Forced to our hands and knees, we climbed obstacles. Descending proved equally tricky, and we sometimes slid and tumbled. Following the patrol, we took our places back in the perimeter. A peculiar smell rose from the ANA dudes to our front.

"You smell that, Ronny B?" Brandon asked.

"Yep, smells like our buddies are tokin' reefer."

"Think we can get some of that?"

"I dunno, man," I said. "You sure you want that shit out here? Bet it's stronger than what we're used to."

"Yeah, you're right."

"But a toke prolly wouldn't hurt."

"Hey, hey, man," Brandon whispered to the ANA dude. "Hashish?"

The stoned little dude motioned for Brandon. We only wanted a sample of Afghan Kush to knock the edge off. Brandon returned with a small piece of hashish. It was a brown Play-Doh-type substance that required burning it with a lighter to break it up.

I pulled a Marlboro from my pack, squeezed some tobacco out, and packed the hashish inside. "Fire it up, B."

He lit it, took a toke, and passed it off.

I puffed. "Ah."

Brandon and I were puffing some of the finest hashish on the planet. Paranoia crept in. The leftover cig was tossed into the bushes.

On the mission's final morning, we stepped off with T-Love on point. For reasons unknown, T-Love had a few brain farts.

I heard Troutman ask Hix as he walked beside me. "Where's T-Love going? What's he doing?"

The lieutenant radioed Stan. "Three-Two, Three-Two. This is Three-Six. What the heck is Three-Two-Bravo doing?"

Stan wasn't pleased. We made our way into the valley, and it was now a simple matter of walking a straight line to Boris, which was in view. T-Love couldn't make it happen, though. In frustration, Stan ordered a halt. We knelt.

An infuriated Iron Guts stormed towards ole T-Love. "Goddamnit, the fuck is wrong with you, T-Love? Why can't you walk a straight fucking line? Aren't you a goddamn E-fucking-Five?"

Stan gestured with his hands, screaming and demanding T-Love get his shit together. "You just wait till we get back, goddamnit!"

While most of us cleaned weapons back at Boris, T-Love got his balls smoked. When someone is getting the shit smoked out of them, it's best to leave the area. Brandon, Joey C, Blackie, and I were out of view but heard it all.

"Man, you gotta be fucked up to get smoked as a sergeant," Joey C said, and we all laughed.

After a half-hour, T-Love returned to the crew. He was breathing heavily and couldn't speak. Sweat dripped from his beet-red face and drenched his uniform. Mud caked his gear from low-crawling through the dirt around Boris. *How humiliating.*

CHAPTER 9: RIDE 'TIL WE DIE

During the early hours of July 4, 2009, members of our sister company, Able Company, endured a fierce attack. Taliban forces opened up on COP Zerok, also located in Paktika Province. Waves of hundreds of insurgents firing rockets, mortars, and small arms ambushed the small outpost. A suicide bomber with an explosives-laden vehicle sped towards the front gate and exploded. Over thirty insurgents were killed. Two Americans were also killed—PFC Casillas, aged nineteen, and PFC Fairbairn, aged twenty. PFC Casillas charged into the enemy fire on three different occasions: once to begin a fire mission, a second time to rescue PFC Fairbairn, who was injured inside a mortar pit, and a third time carrying PFC Fairbairn to the aid station. While Casillas rushed Fairbairn to treatment, a mortar impacted a few feet away, killing them both.

The Bible says, "Greater love hath no man than this, that a man lay down his life for his friends." The Independence Day attack will forever be ingrained in the minds of survivors. *Geronimo.*

News of the attack served as a harrowing reminder. The ghosts we're chasing are real. When boredom takes over, complacency sets in. One thing

Stan preached was "never become complacent." As odd as it seems, it's tough battling complacency. But the events of July 4, 2009, put us on notice.

Now that we operated from the FOB, we had far-reaching missions that included all regions of the AO. Sometimes we escorted civil affairs officers to schools to meet with local teachers and discuss methods to improve education. We drove civilians to reservoirs to test the water, for God knows what. Shuras were conducted. Security patrols pushed down wadis and through villages. We were often tasked with route clearance, which is an oblique way of saying drive a certain road before more important people drive the route later. Perhaps a high-ranking officer, politician, celebrity, or other VIP was passing through. Third Platoon could trigger an ambush beforehand. We were the canary in the coalmine.

The most hated missions were known as Battle Damage Assessments (BDA). A BDA followed rocket or mortar attacks. The platoon mounted the MRAPs and convoyed straight where the enemy rockets were launched. The problem with BDAs was the simple pattern – the enemy launched rockets, and we drove to the POO site. It's predictable, so we'd walk into an ambush. Word was that the brigade commander required BDAs after every attack to ensure civilians weren't harmed by counter-fire artillery. But the perpetrators were rarely found because of the time it took to reach POO sites. The Taliban was masterful in recovering dead bodies quickly and blending amongst villagers.

We had one important mission during this time, providing security for the Afghan Presidential Election. We ensured locals could get out and vote without fear of intimidation by the Taliban or others who hated democracy. Third Platoon drove to the northwest of Boris and provided overwatch high atop the mountains. The ANA set up at the polling stations. They needed to be visible, providing security for their nation's election. We operated in the shadows. A couple days before the election, we camped in the hills and patrolled the area, deterring and disrupting enemy activities. On election day, we were tucked in the mountains, ready to strike if drama ensued.

We did our part to provide a safe atmosphere for citizens to vote, but the Afghan government failed to deliver. The election was tarnished with corruption, intimidation, and ballot stuffing. Incumbent President Hamid Karzai was declared the winner on August 20, 2009, with almost fifty percent of the vote. His challenger, Abdullah Abdullah, garnered a bit over thirty percent. American puppet Karzai wouldn't be defeated. Others said this election day was the most violent in Afghanistan since 2001. Our AO was calm other than a few rockets blasting near our position. The people in the district were generally uninterested in the government in Kabul. A lack of interest also indicated a high Taliban or Taliban-sympathizing population.

Overall, the election was a failure. Voter turnout was low, and corruption was high. The American and Afghan governments wanted us to believe Afghanistan is a democracy. It wasn't. A runoff election was scheduled for November 7, 2009, but Abdullah Abdullah refused to participate. Hamid Karzai was once again declared the winner by default.

After a few days in the bush, I showered and relaxed in the rear. The Dog boys hosted small parties, and I sometimes stopped by. Incense burned in the doorway and something else burned inside the room. Movies, music, and video games played while guys indulged in things that lifted them from this strange world for a moment. I wasn't interested in the hard stuff, but I never minded a toke from the peace pipe to ease my mind.

Afterwards, I'd belt out a little Flair. "You might not like me, but you'll learn to love me because I'm the best thing going today. Wooooo!!"

The Dog boys followed. "Wooooo!" We hollered, sang old country lyrics, and laughed away the night.

Inside the MWR, I'd log on to the internet and chat with my sister or available friends. I hadn't spoken to many of these people in fifteen years. But thanks to Facebook, they were easy to locate. People were shocked at my current situation. Ole Ronny B was knee-deep in war. Most of my generation who'd served did so during peacetime nineties, as I did. Even former high school students tracked me down. Their high school history teacher was part of the war effort. Inside the cyber world, I was a curiosity. Old acquaintances sought me to offer support, ask questions, pray, and scroll my profile.

Ladies popped up in my chat bar. Some were from Alaska or even further in the past. Their messages were sometimes bold and steamy, evolving to sexts. Some gals were married with children. One married gal sent hot nude photos and offered a sexy rendezvous when I returned to Atlanta. I never initiated conversations with married women, but I'd oblige if they wanted to chat. They loved to roleplay. The ladies in the world viewed me as a hero even though I was only an older Joe near the bottom of the totem pole. Folks love playing make-believe with people they can't have.

The talk was fun, but I wasn't really interested. Jane popped up from time to time. She sent some photos of her new breast implants and her dressed in a skimpy police costume. It was a morale booster.

Hanging at the Dog house or online is fun, but soon it's time to get back to business.

While returning to base after a patrol through a bazaar, enemy rockets peppered Boris from Rocket Ridge. Artillery from Boris blasted back.

"Fuck my life," Joey C said. "We're going back out."

We made our way to the "cans" to await further instructions, and I encountered a stunned Blackie. He wasn't on our last patrol.

"Dude, you won't believe this shit," Blackie said with vigor. "I almost got smoked by a goddamn rocket!"

While Blackie sat by a Hesco wall smoking a cig, a rocket ripped into the opposite side and jolted the ground.

"We gotta roll," Stan said and rallied the squad.

The brigade commander ordered us to be at the POO site within an hour, but the nearby MRAP had four flat tires. Anyone can change a tire but changing four massive tractor-style MRAP tires is a bitch. It was late afternoon before we were ready to roll. The brigade commander's timeline wouldn't be met. Daylight was fading, and everyone felt ill over the BDA mission. After arriving at the base of Rocket Ridge, others dismounted and pushed out with night vision gear. Smitty and I stayed put in the vehicle inside a hasty perimeter with three other MRAPs.

The Afghan sky was pitch black when the dismounts returned. Lieutenant Troutman, Manny Fresh, and the squad leaders held a quick meeting to discuss our return to base. Manny decided, against the squad leaders' wishes, to drive back blackout style – headlights off and drivers wearing NVGs. None of the drivers had experience driving with night vision equipment, though. The squad leaders pleaded with Manny and the lieutenant not to go blackout. The route back contained major dips and steep drop-offs. Stan chimed in that it'd be a good idea to train everyone to drive blackout on the FOB at least once before doing it "outside wire." Manny refused to budge, and Troutman backed the platoon sergeant he knew little about. Although Troutman yielded to his platoon sergeant this time, it would be the last time.

After the meeting, Stan pulled Manny to the side to plead his case one last time.

Manny scoffed. "Yuz juz wonna getz back to zee FOB as wick is pooseble so fook off."

Total darkness blanketed the sky when we drove off. As the gunner, I was the eyes and ears for the vehicle, sitting up higher than everyone else. In a normal situation, I'd scan for wires, suspicious vehicles, and enemy fighters. But now, I was reduced to aiding Smitty in driving. Stan rode shotgun as TC (Truck Commander) and helped with directions. "Over to the right, over to the left," I'd say, doing my best to help keep Smitty on the road.

The MRAP dipped too far left into a steep drop-off, and the road started to give way. The vehicle began a slow-motion roll towards its left. The ground beneath the left-side tires collapsed as the right-side tires rose from the dirt. *Oh shit.* I clutched the Mark 19 handles, lowered my body into the tuck position, and prepared for impact. Over fifteen tons of armored steel came crashing down.

I was flung like a rag doll and landed on my back. My boots swung overhead, folding my pummeled body like a pretzel. My ass pointed upward. Mark 19 ammo cans, strapped across from me, broke loose and slammed my backside. Mark 19 rounds are the biggest, heaviest, and bulkiest. The weapon itself broke loose from the mount and buried itself into my chest. I was thankful for my armored chest plate. Thankfully, the MRAP stopped short of flipping upside-down, or I would've been toast. I patted my ass and moved my hands downward towards my feet. *It's all there.*

Intense yelling rang from the cabin. "Bruce, Bruce!" Stan went on. "Answer me, Bruce!"

Still lying feet overhead and ass poking up, I mustered as much voice as I could. "I'm okay, sergeant."

"C'mon Bruce, answer me!"

"I'm fine, sergeant."

"Hold tight, buddy, we're gonna get you outta there." Stan sounded relieved.

Guys from other vehicles headed my way. Doc Rodrigo was first to arrive, and he slid in feet first, baseball style. "Hey, you okay?"

Once Doc confirmed I was safe to move, he single-handedly dragged me from beneath the vehicle, pulling on shoulder straps of my body armor. Concussed from the impact, I laughed.

Other members from the platoon, led by Sergeants Donald and Rubio, attempted to rescue everyone still trapped inside MRAP – which included Smitty, Stan, Hix, Stu, Phil, and Nate. I could hear their screams from within the vehicle. The heavy armored door to the rear was difficult to open since its hinges were on the right side. The weight of the door had to be pulled upwards. Stan was soaked in fuel from the vehicle's leaking tank, so he ordered Smitty to shut down the power. So, the pneumatic system that typically opens the rear door didn't have air pressure. Finally, the top hatches were opened, and everyone was pulled from the wreckage. Sergeant Fink rallied the stunned platoon and placed everyone into a defensive position as leadership mulled over the near disaster. We needed to hold the line as Dog Platoon spun up to help recover the wrecked MRAP.

Are the Dog boys high as fuck? Imagine a late night, watching a movie, smoking hashish then boom, QRF. Stoned or not, it didn't take long for Dog to arrive. The heavy-duty wrecker snatched fifteen tons of wrecked steel and towed it home. We were placed in a difficult situation but came together and used teamwork to overcome it. I knew this platoon would never quit nor leave a man behind. *Dog's got us too.*

We arrived at Boris, and all of us that were inside the crashed MRAP headed to the aid station for a brief examination. The first sergeant looked at me with a smile. "Aren't you glad you wore the harness, Bruce?"

"Yes, first sergeant."

"Glad you're okay, son."

The first sergeant was hinting at his new policy that required all MRAP gunners to be strapped in. It was the first time I'd worn a harness, and it may have saved my life.

I later found Doc Rodrigo and thanked him for dragging me from beneath the vehicle. "That's my job, man." He said.

"Yeah, but you were on point, baby."

Despite being battered, I told Doc I was okay. I wasn't leaving the fight. A little pain is nothing a grunt can't handle. I'd get evaluated following deployment and deal with medical issues back in Alaska.

Doc Rodrigo was soon replaced by a new medic. Rodrigo had done nothing wrong; more young medics needed time on the line. Doc Lindley, a light-skinned, blonde dude from Minnesota, was the new addition. He dove in headfirst and blended well with the guys.

After our medical evaluations, Stan drifted off for a confrontation with Manny. It was Stan's "I told you so" moment. They cursed at one other. Manny threatened Stan with an Article 15, and Stan clapped back that "an ass-whooping was coming."

"Take my goddamn stripes, motherfucker!" Stan told Manny.

His men meant more than his stripes. But Stan continued to beat himself up over the rollover incident as if he could've done more to prevent it.

On a windy Boris night, Manny Fresh gathered the platoon to explain a mission the following day. Manny said his piece, but no one understood what he said. He left without further explanation.

Hix stood and said. "Fuck that, here's what's goin' down."

"We're gonna drive to Margah on Route Volkswagen and pick up a few dudes from Second Platoon. We'll drive them back to Boris, so they can jump on a bird and make their way to Bagram to fly out for R&R. Why the goddamn bird can't fly to Margah to scoop them up is beyond me. We're gonna play chauffeur. Also, fellas, on way home, we're gonna stop along Volkswagen and search a few houses. Some fucking intel reports indicate that shit stain DUSTWUN kid, Bergdahl, may be held in the area. Watch your asses and gunners be alert because Margah is a bad place, and Volkswagen is a dangerous route. Lotta bad dudes there. Get some sleep and be ready in the am."

As the sun peeked above mountains, Third Platoon readied the vehicles to drive to Margah. Gunners mounted weapons, checked headspace and timing for the .50 cals, and strapped in their harnesses. We performed radio checks and double and triple-checked the gear loaded inside the MRAPs.

Lieutenant Troutman gathered the platoon to give final instructions and lead us in prayer.

"Guys, keep your heads on a swivel and use situational awareness," Troutman explained. "Communicate with one another and don't hesitate to report anything suspicious."

It was now prayer time, so the lieutenant flicked his cigarette to the ground and prayed to Jesus. "Dear Heavenly Father, we ask that you bless us on this mission . . ."

As Troutman prayed, I peeked around at the boys in curiosity. Whether prayer took place at family gatherings, inside the locker room before a high school football game or in church, I always opened my eyes and let them wander. I noticed best buds, Stan and Hix, two nonbelievers, staring straight ahead. But they did so with respect.

"In Jesus' name, we pray. Amen," the lieutenant concluded.

During the buildup to each mission, I felt like I was inside a football locker room before a big game. The speech, the prayer, the high stakes, the men, the intensity, the rush: it was all there. Before most missions, I'd go to MWR, log on YouTube, and watch Derrick Moore deliver an impassioned speech to my beloved Georgia Tech Yellow Jackets.

Our four-vehicle convoy wiggled through the serpentine (winding barrier wall) to exit the FOB. I sat in the back of an MRAP while Brandon drove, Stan was TC, and Joey C gunned a .50 cal. The wadi was wide on our route. Neighborhoods of kalats sat to one side. Tall cornfields surrounded both sides of wadi along the entire route. It was difficult to see what was hidden inside those massive cornfields.

We arrived at Margah and parked the MRAPs at the base of the hill leading to COP. Dismounts pushed out a patrol. Rubio led the patrol, and I followed him as we looked around homes and into yards. Rusty and I positioned ourselves at the corner of a kalat, and locals gathered at a distance.

The villagers didn't appear happy to see us. Hate emanated from their faces. Ladies were out: hanging clothes, gathering buckets of water, and performing household labor. I smiled and waved at the women, and Rusty shook his hips and performed a dance. The ladies, covered in burkas, ignored us. Pashtun culture is next-level misogynistic. Women aren't allowed to look a man in the eyes unless he's the hubs. Speaking to an American male is a violation that can result in beatings or death.

The convoy mounted for phase two of the mission—look for Bergdahl. We had no idea where intelligence originated or if it was credible. But we'd roll back down and search the designated homes.

A Second Platoon dude heading out on R&R boarded our vehicle. I looked at him with a smile and said. "Must be nice, man. I'm happy for you."

He was Second Platoon's medic and replied with a short "thanks," as he sat looking nervous.

Perhaps he felt he wasn't out of the woods yet. He understood a trip down Route Volkswagen stood between him and the freedom bird. Second Platoon had been ambushed along this route many times. The young man understood what we from Third Platoon were oblivious to.

We reached the area of target homes, and Manny Fresh ordered the convoy to split. This led to another disagreement between Manny and the squad leaders. Manny and Rubio's MRAPs drove around the corner into a neighborhood to dismount and search. The other two vehicles, led by Hix and Stan, stayed put inside the wadi. To my dismay, the guys conducting the search were out of our view. Time ticked by. Stan and Hix chatted over the radio, asking "what the fuck was taking so long?" One hour, two hours, three. Everyone grew nervous. Stan called for a SITREP (Situational Report).

A colossal explosion erupted. I glanced through the back window of the MRAP and noticed the vehicle to our rear had been rocked by an RPG blast. Flames and gray smoke drifted upwards.

"They're hit!" I screamed as madness ensued.

RPGs and small arms rounds were unleashed from the thick cornfields. The enemy smashed our vehicle with a fully cocked sucker punch. Dirt kicked from the wadi below as bullets from AK 47s impacted around us. Another RPG pounded the wadi just meters from the rear of our vehicle. I located the enemy origins of fire due to the massive plume of smoke rising from RPGs after they're fired. It was time to fight back.

"Point that fifty to your eight o'clock and fire that bitch!" I yelled to Joey C.

"Fire Joey; light those motherfuckers up!" Stan commanded.

Joey C fired away into the cornfields. I again peeked through the back window with concern for our hit vehicle. It was still reeling from the explosion. Another RPG detonated onto the vehicle, blowing the sniper cover off the turret. Gus, the vehicle's gunner, dropped into the bottom of MRAP.

"Sergeant, they're hit again," I screamed.

I thought Gus was dead. For a moment, I remembered the blonde and balding guy from Wyoming drinking "near beers" during chow and smoking Marlboros afterward. But he wasn't dead. Gus rose inside the turret and laid down well-placed Mark 19 grenades into the cornfields; it was unquestionable bravery on his part. Wounded and bleeding, he stood his ground versus a well-organized enemy. To avoid sitting like ducks, Stan ordered Brandon to drive the vehicle in circles as Joey C blasted away with the .50 cal.

The other element of the convoy appeared. Sergeant Rubio sprinted alongside an MRAP to make his way to the hit vehicle.

Relieved, I yelled to Stan. "Sergeant, we gotta medic back here. We need to help those guys!"

"Go!"

I approached the smoking vehicle and tried to open the driver's side door. But I couldn't get the warped armor to budge. I sprinted to the rear of the

MRAP to get into the backdoor. As it opened, Rubio, Stan, and Doc Lindley, succeeded in extracting KG from the MRAP's driver door. A shaken Hix, Jordy, Gus, and Trubisky exited through the rear. I darted back to the MRAP's front to see what was going on with KG. Blood bubbled and foamed on his mouth. His Kevlar helmet was ripped to shreds. The back of it resembled shredded cheese.

Doc Lindley worked on KG calmly. "Stay with us buddy; you're doing fine." Blood gushed, and KG's eyes rolled back.

I didn't think KG would make it. He'd taken a direct RPG explosion to the dome. The firing ceased as we treated our wounded, and the enemy retreated. We needed to get the battered MRAP moving. Gus, Jordy, and Hix were wounded yet not critical. KG was in and out of consciousness. Hix stumbled around with a confused look on his face because he'd suffered a concussion. KG had absorbed the brunt of the blast while the remaining shrapnel peppered Hix as he sat shotgun.

Hix attempted to cut a stretcher loose from the MRAP, so I approached and said. "It's okay, sarge; we don't need it."

Hix wanted to lead, but his mind wasn't putting two and two together. Stan stepped forward to look after his best friend, so I knelt by the front corner of MRAP and scanned cornfields through my optics. Another ambush was possible as we struggled to get the MRAP running.

The disabled MRAP soon fired up, and we raced against time to evacuate the wounded. The area where the firefight took place was too hot for a helicopter to land, but birds were on station when we arrived at Margah. Gus, Jordy, and Hix stumbled to the helicopter while KG was carried on a stretcher. *God bless y'all.*

A perimeter was set up at the base of COP Margah while we planned our next move. The medic who was supposed to be heading home moped as he walked back up the hill. While we were dusting off the stains of battle, a small vehicle sped towards our perimeter. The driver frantically waved a white flag. No one fired, and the vehicle crept near our perimeter. The driver stepped out, walked to the passenger side, and opened the door. He reached inside and pulled the small body of a seven or eight-year-old boy from the vehicle. While tears streamed down the poor man's face, he forced his way to the center of the perimeter and knelt before Stan and Manny, grasping his deceased son's limp body.

I glanced at Rusty, and Joey C. None of us said a word. Doc Lindley examined the boy's body and concluded the fatal shot was caused by small arms. Manny produced a .50 cal and Mark 19 round in a fruitless attempt to persuade the man that his son couldn't have been killed by us. The bulky rounds would cause much more damage. But there was little to be done to console the man as he wept over his son. I eyeballed the grieving man when he carried his son back to his car and drove off into the wadi. The image of the dead child lying in his father's arms was burned deep, forever ingrained in my mind.

Dog Platoon convoyed to Margah. Stan left with them to perform a BDA where the firefight took place. The rest of us from Third Platoon mounted vehicles to return to Boris. Sergeant Rubio approached and asked if I'd gun the hit vehicle. I was honored to do so. Smeed drove, and Rubio served as TC. I strapped in the harness and entered the bloody vehicle. The packs and gear of the wounded remained inside. Globs of blood covered their gear. I stood tall behind the damaged Mark 19 with a SAW posted for backup. As we rolled through the scene of the ambush, I kept my finger near the SAW trigger.

We returned to the FOB, and dudes who weren't on a mission awaited us. They rushed our vehicles to break down gear, remove weapons from mounts, dispose of expended rounds, and scrub blood from inside the vehicle. I sat on a bench outside my room, fired up a cig, and reflected on this hectic day. I didn't visit the Dog House. I remained alone and tried to make sense of everything.

Later, I approached Joey C as he chilled on a bench outside his room. "I'm proud of you, buddy. You were a rock star today."

"Appreciate it, Ronny B," Joey C said and toked on a smoke. "Man, what a day."

We later learned that KG made it. He'd be sent home following a lengthy stint inside a hospital. I didn't know the extent of his injuries, but it was passed down that he suffered a traumatic brain injury. Word was that he lost a significant amount of vision and hearing. But he still had his lucky Garfield, which had been stuffed inside his pocket. He had carried it everywhere. Jordy suffered damage to his eyes and was sent back to Alaska to recuperate. He made a full recovery and later served on another deployment. Gus and Hix suffered shrapnel wounds and concussions. They were sent back to the platoon after being patched up at Bagram.

Following the ambush, it was rumored the civil affairs sergeant on the FOB was critical of our response. She questioned why we fired into cornfields and blamed us for the death of the child. What a low blow. Machine gunners are taught to return fire as fast as possible to either force enemy's heads down, lead them to retreat, or kill them. It's known as suppressive fire. None of us cared to be lectured by a "fobbit" (slang term for one who never leaves the relative safety of FOB). The situation was tough enough without her judgment and condemnation.

A couple of days after the firefight, the platoon was summoned for our next mission. We'd be traveling back to Margah and driving Route Volkswagen to escort a drug tester to the COP. That "meat gazer" (someone who looks at your meat while you piss) would administer a urinalysis to Second Platoon.

"We're driving out there for what?" I asked Joey C. "You gotta be shittin' me."

The higher command didn't want us to know the reason for the mission because they knew we'd be irate.

But Manny Fresh let the cat out of the bag when he met with the platoon to discuss the plan. "Weez goona go oot zere zoo geev zem a gozam peez tez. Itz boolsheet but weez goona zoo it."

The powers-that-be felt Second Platoon had a drug problem. Rumors suggested that drugs were obtained through the mail and dispersed amongst some platoon members. It was also suggested that pills and dope were bought from a pharmacy inside the Margah bazaar whenever squads patrolled that area. Whatever the case, Third Platoon was being dragged in.

Late that night, several of us were getting our workout on. As I stood by the dumbbell rack, gripping weights, and performing shoulder shrugs, I had an eerie conversation with Ross.

"Whatchu think about this bullshit piss test?" I asked.

"Fucking shitty, man," Ross replied.

"Wouldn't it be fucked up if one of us got hit during this mission?"

"Dude, that'd be shitty."

In the am, I woke up and strolled to the chow hall for breakfast. Eggs over easy and strips of bacon piled my plate. Breakfast was my favorite meal on Boris. I then headed to the MWR to log on to the internet and send a message for my sister to pass on to my father. I dialed up the usual Georgia Tech football pre-game speech on YouTube to get jacked. Afterward, I caught up with the platoon. Lieutenant Troutman gave his speech and led us in prayer. I peeked at Stan as he stared off into the distance. My excitement returned as I banished my thoughts about the purpose of this mission.

The ride down Volkswagen was smooth. We parked at the base of COP Margah while the meat-gazing "fobbit" did his thing. *Are people nervous about pissing inside a cup? Were they warned?* The samples were collected, and we mounted up to return to base. Near the exact location of the previous ambush, all hell broke loose, and fireworks popped off. The enemy was in the cornfields again.

"Light 'em up, Dottie!" Stan screamed.

Guns fired from every vehicle as we raced down Volkswagen. This time we were on the move as opposed to sitting still. The rounds from the vehicles penetrated adjacent homes, ripped through cornfields, and scattered crowds of people.

I noticed smoke rise from the enemy's RPG fire, and I screamed at Dottie, "Shoot! You see those motherfuckers!"

Dottie mashed the hammer down. I glanced through the back window and watched Phil rip his .240 and lay scunion on those bastards. Up ahead, ole redneck D-Rum was behind a .50 cal. We drove through rockets and small arms fire for nearly a mile down the route.

A well-placed RPG struck the top of our vehicle. Dottie crashed downward through the turret and into the bottom of MRAP. *Fuck!* Smitty, Rusty, and I reached to grab Dottie to give him medical aid and get him out of the way so one of us could leap into the turret and continue firing.

"Dottie, you alright?" Stan hollered.

In an instant, Dottie rose from the bottom and grabbed the machinegun. He continued firing like a man on a mission.

"Goddamn right, Dottie!" Stan said.

When we neared the FOB, the firefight fizzled out. Once the vehicle parked, Stan inspected the damage.

"Goddamnit, you're a lucky motherfucker, Dottie."

The bulletproof glass was blown from the turret, and the armored steel was twisted and warped. Dottie didn't have a scratch on him, nor did he receive a concussion from the blast.

The occupants of the vehicle to our front weren't so lucky. RPG rounds struck the MRAP and riddled guys with shrapnel. Smeed, Nate, and Ross were rushed to the aid station and later to a hospital. I reflected on the conversation I had with Ross inside the gym. It was prophetic. Nate and Smeed wouldn't return, but Ross would be back in a few days. The meat gazer and piss samples were inside the hit vehicle. I'm not sure what happened to the piss, but the meat gazer was unharmed. This incident will forever be known as the "piss test ambush." *Maybe Second Platoon caught a break?*

The platoon was down seven men, and we didn't have the required manpower to roll "outside wire." We needed a rotation of force protection. It was at this point I grew disillusioned with the war effort. In a three-day span, we'd gotten seven men wounded while trying to chauffer soldiers to R&R, search for a deserter, and administer a piss test. We were getting picked apart by guerillas from the Stone Age. I didn't believe these fighters had the means, ability, nor desire to do damage inside my own homeland. They fought us because we were in their backyard. It was becoming a turd sandwich. I made up my mind that I was doing this for the men around me—not the bullshit greater good.

CHAPTER 10: THE BEAT GOES ON

Pulling guard with Afghan dudes from the ASG was interesting. They knew little English but understood buzz words they'd picked up from young soldiers who gave them an education in smut. Their responses went as follows: if they dislike something, it's a "no," if they like it, it's a "good," if they love it, it's a "double good."

During one shift, I asked the thirty-something Afghan. "Do you like pussy?"

"No."

"C'mon, man, you don't like pussy?"

"Okay," the man said with a nod.

"Do you eat pussy?" I laughed and made a pussy-eating gesture with my tongue and fingers.

"No."

"Do you like ass?"

"Ass good."

The Afghan fixed me with his dark eyes and said, "Look, man with woman, okay. Man with man, good." He paused and gestured his hand low to display a child's height as his face lit up, "Big man with little man, double good!"

"You sick fuck!" I blurted, which he didn't understand.

As shocking as it is, the molestation of boys is a cultural practice in Pashtun culture and other Afghan societies. The conversation made me remember patrols where I witnessed preteen and teenage boys wearing mascara to look feminine. The practice of exploiting boys is known as "bacha bazi" or "boy play," and it's seeded deep in Afghan culture. Phil reminisced about a patrol where we walked up on a nervous-looking grown man lying on a blanket with a frightened young boy. *Busted.*

The sad fact is that child molestation in Afghanistan is rarely punished. It's a cultural norm to many people. Sure, pedophilia exists in the US and Western culture. But in Western societies, people convicted of molesting children are thrown in prison. Once in prison, the molester is a marked man. You know it's a serious offense when the molester is deemed a deviant and outcast amongst convicts. However, a man getting his "bacha bazi" on in Afghanistan is given a high five rather than a beatdown.

Women in Pashtun culture don't fare so well, either. They're severely limited in their freedoms. They dress in burkas with only their eyes exposed every single day. Ladies are subjected to tons of abuse for minor infractions, at least minor by my Western standards. Causes for punishment include speaking to men they're not married to, appearing outside without a burka, or seeking education or employment. Women are beaten in public, mutilated, and acid-bathed for things considered trivial to average Americans.

I picked the brains of our interpreters because I was curious about the culture. One interpreter told me about his three wives.

I asked. "What do you do if one of your ladies gets outta line?"

With a smile, he made a backhand motion and smacked the back of his hand against his other palm.

"You smack a bitch if she gets lippy?"

"Yes," he said with a nod.

We both laughed. I'm sure the man understood certain sectors of western culture glorify hitting women who act chippie. In practice, though, men putting hands on a woman is unacceptable in most Western societies. Nowadays, it's considered shameful to hit a woman, as it should be.

I was well-acquainted with most ASG guys since I'd sat and bullshitted with them on guard duty. There was rarely drama. However, things got heated up one night after I noticed an unfamiliar Afghan on the shift. I leaned back with my NVGs mounted, propped my weapon against the wall, and hunkered in for what I hoped to be an uneventful night. The conversation with the new guy turned into the usual "do you like pussy?" But this guy was loud and obnoxious. I rolled with the punches and pretended to listen.

The Afghan went on about how he loved ass, meaning dudes, and asked, "Do you like ass?"

"No," I replied.

"Why?" the man asked over and over.

"Ain't how I roll."

Then, as I stood behind .50 cal attempting to scan the area, the man grabbed my crotch.

"What the fuck?" I ripped his hand from my crotch and twisted his wrist. He lost his balance and crashed into the chair.

"No touch!" I said, raising my open hand. "Touch me again, and I'll slap the fuck outta you, boy."

I grabbed my rifle, and the man attempted to ask in English what was wrong. "No fucking touch," I repeated.

The other Afghan sat in silence.

During the remaining three hours, we sat in the darkness, stone silent. I clutched my weapon in my lap and kept the corner of my eyes glued to the two Afghans armed with AK 47s. Afterward, I heard the guys' usual "I would've beat his ass" when I mentioned the incident. But I knew I'd gotten my point across without having a nighttime shootout that could have resulted in death. I had the drop on 'em but didn't need the drama.

"Time to get up," Stan said as he entered our room and hit the lights. "Be at HLZ in thirty; it's air assault time."

A glance at my watch showed it was 0200, so I rushed to get dressed. Stan ensured that we kept our assault packs stuffed with MREs, cold weather gear, and everything it took to spend at least three days in the bush. It was obvious ole Iron Guts relished the thought of a potential high-speed mission. Our platoon wasn't at full-strength, but it didn't need to be because we were supplementing Captain Mac and his headquarters element.

"Waz up, jarhead?" I heard that smooth, Mathew McConaughey southern tongue.

"Sergeant Dutch, waz up, baby?"

He was geared up and ready to go. "Alright, alright, alright; let's get some."

While awaiting the arrival of the bird, T-Love scrambled around, trying to stuff his pack with knee and elbow pads. Stu approached and shouted. "What the fuck, T-love? Whatchu need those for? You going skating?"

I almost spit my MRE cracker out laughing. T-Love handed his knee and elbow pads to Stu as everyone chuckled. Stu was heading out on R&R soon, but he still ensured his guys were ready for the mission. Once the bird arrived, we loaded and prepared for our arrival to no man's land. I looked across at Blackie as he flashed the horns with his fingers \m/. Joey C smiled like he was excited, and Iron Guts was pumped because this was what he lived for.

"Not everyone gets to do shit like this, Bruce," Rubio said as he sat beside me.

The bird landed under the cover of darkness. Everyone sprinted to the ground and set up a defensive position. Captain Mac and his entourage linked with the special forces men who'd raided the area earlier. These guys were the elite of the elite, bad motherfuckers, and when they went to an area, they didn't fuck around. I referred to them as "long beards," and local villagers called them the bearded soldiers. They operate with impunity without the politically correct restraints that hold back conventional forces.

The "long beards" had raided a village and killed several people. It was up to us to deal with the aftermath. I referred to these air assaults as "special forces cleanup missions" since we were tasked with repairing relations with the locals and disposing of dead bodies. The "long beards" took off after our arrival.

A nearby house hadn't been cleared yet, so Second Squad was ordered to clear it. We snuck to the kalat and stacked outside the door. T-Love would be the first to enter. We busted inside, hoping to be fast and furious, but ole T-Love had a brain fart. He froze inside the lethal "fatal funnel" of the room and clogged everyone up until someone rammed him from behind. The home was empty. Stan scolded T-Love for a mistake that could've been deadly in another situation. An assaulting force is most vulnerable while inside the fatal funnel of a room.

Once the area was secured, the cleanup portion of the mission kicked off. Guys pulled dead bodies from the carnage. Some were bad guys, and some weren't. This enemy was known to hide behind innocent civilians. Sometimes they'd takeover homes of peaceful villagers in entire neighborhoods and stage ambushes from the kalats. Once the cleanup was completed, the birds returned and scooped us up. We made it back to Boris in time for lunch.

One of the more memorable air assaults followed the dropping of a couple five-hundred-pound bombs onto a gathering of Taliban. Rather than extracting bodies, we happened upon bits and pieces of body parts dispersed along the patrol. A foot here, finger or toe there. The Taliban is proficient in recovering bodies, but they were unable to police each chunk of decimated remains. When millions of shreds of shrapnel explode from above, the damage is horrific. Blackie enjoyed the show. He kept flashing the horns with his fist and saying how gnarly it was.

"Wake up," Stan said. "We're doing another air assault."

It was 0300 this time. Another special forces cleanup mission was handed down. A bird arrived and flew us to a field close to a large neighborhood. We landed and set up a hasty perimeter. I watched "long beards" move towards the helicopters with a line of prisoners they'd rounded up. I counted fifteen prisoners—tied together with rope, duct tape covering their eyes.

We moved into the "hood," and the platoon took over a target house that'd been raided. Stan pushed his squad onto the roof to provide overwatch while the rest of platoon took cover. I rolled with Captain Mac and Manny Fresh with the HIDES device in tow. HIDES gathered biometrical data by scanning pupils and lifting fingerprints. It would snap pictures and store personal information inside a database. Manny Fresh rounded up a group of military-aged men and lined them up. One by one, I scanned each pupil, extracted a fingerprint, and snapped a picture.

Once the HIDES work was completed, Manny instructed me to link with Stan on the rooftop. When I entered the home and was alone inside, I decided to give myself a tour. I wandered into the living area and inspected the Persian rugs and clay pottery. There were wooden bookshelves filled with books and other items of value. The walls were covered with yellow and red wallpaper with a plaid-like design. Photographs of the countryside and pictures of what appeared to be family members hung from the wall. Some chai teapots caught my eye. I stuffed some into my bag, along with a Haji scarf that hung from a shelf. No one would've cared that I collected some spoils of war, but I thought better of it, and I returned all the items.

In the bedroom, I made a gruesome discovery. There was a bearded Haji lying on the floor. His wide-open eyes stared into mine. Blood soaked his white garment from his neck to groin area. He'd been gut-shot. For a moment, I examined the man in complete silence. We looked at one another. *Is he meeting Allah or dead to the universe?*

I made my way to the rooftop and spotted Stan examining the neighborhood. I grabbed his attention.

"Waz up, Bruce?"

"Dude, you gotta see this shit."

Stan followed me down a stairwell and into the room where the Haji was taking a permanent nap. "Goddamn, they shot the fuck outta him."

Stan radioed Captain Mac, and he entered the room with Fink. The captain glanced over the dead Haji. "How 'bout getting some biometrics from him?"

I snapped pictures of the man's pupils while Fink forced his eyelids open with his fingers. We lifted his hands and placed each finger on the HIDES to capture prints.

While straddling over the man and gathering info, I heard a whisper from behind. "Bruce, turn around and stick your thumb out."

Without a thought, I spun my upper body and stuck my thumb out. Fink snapped a pic with his personal camera. I had a shit-eating grin on my face. This picture isn't for public consumption, though. Modern politically-correct society would excoriate me over it. *What happens in combat stays in combat.*

CHAPTER 11: THE MADNESS OF MARGAH

The platoon soon learned that Margah was the wild west, and a drive on Route Volkswagen was almost guaranteed to trigger an ambush. The only other option was a more scenic route east of Spaghetti Hill. The alternate route was more time-consuming but safer. The enemy had fewer opportunities to stage ambushes because the surrounding area was wide-open with no vegetation. The cornfields aligning the wadi along Volkswagen were a bitch to us. When we discussed them, Frankie mentioned. "If we played by the same rules we used in Iraq, we would've torched those cornfields after the first ambush."

He had a point. The counterinsurgency rules of engagement (ROE) in Afghanistan were more restrictive than they were during Frankie's time in Iraq.

On a bright sunny day, Stan summoned the squad. "I think you'll like this, fellas. The order to teardown COP Margah has come down."

Holy shit.

"The brass has decided COP Margah is no longer needed," Stan said.

"Fuckin' A." *Best news I'd heard in a while.*

"Soon, we won't have to worry about that shithole anymore," Joey C said, and he took a drag from his cig.

Word was the high command wanted to hand over COP Margah to the Afghan Border Police (ABP). But the ABP didn't want it. So, someone way up the chain decided to destroy the COP.

There was a half-built ABP compound located at the base of the hill where COP Margah stood. The ABP compound sat in an undesirable location on the low grounds. It had target practice written all over it. Several enemy-occupied mountains and well-known indirect fire POO sites surrounded the compound. The buildings were brittle, and the infrastructure was weak.

Third Platoon mustered inside the conference room to get the scoop. Manny Fresh stood, offered brief instructions, and left the room.

When the door slammed shut, Stan stood and said, "Fuck that, forget everything you just heard."

Rubio joined Stan and took over the room. "This is how it's gonna go down." He explained the mission and pointed out our movements on large maps sitting on the conference room table.

On this ambitious mission, we'd join Captain Mac and his Headquarters entourage, Dog Platoon, and an element from the Georgia National Guard. We never interacted with the guard dudes before. A brigade photographer would also tag along to document the mission; this was the only time an outsider with a camera accompanied us.

We'd convoy to COP Margah in two elements, driving two different routes. Third Platoon, Dog Platoon, Headquarters, and a handful from the Georgia National Guard would use the east route. The Georgia Guard element, aka Vampire, would drive Route Volkswagen with an ANA platoon. The guard guys volunteered to take Volkswagen because they wanted to "get some." *Okay then.* We'd meet at COP Margah, and Second Platoon would kick off teardown.

For three days, our massive element would patrol and attempt to clear the wadis, cornfields, villages, neighborhoods, and homes of Margah. At the mission's conclusion, Dog Platoon would remain in Margah to assist Second with the teardown. The rest of us would return to Boris and relieve a tiny skeleton crew tasked with force protection. Two weeks later, Third Platoon would roll back to Margah, relieve Dog, and assist Second in completing teardown.

The convoy readied itself to roll, and I took a seat in the back of MRAP. I'd serve as a dismount. Our platoon was still missing guys due to injuries. We had a reservist civil affairs staff sergeant gunning, and a Georgia National Guard sergeant included as a dismount. It's unusual for a staff sergeant, much less a reservist non-infantry one, to serve as a gunner. Sergeant Bess heard our platoon was down a guy and volunteered for the gig.

Since the guardsman served the Georgia National Guard, I said. "I'm from Georgia, myself."

"Oh yeah?"

"Yep, Jonesboro, just south of Atlanta."

"I'll be damned," the guardsman said. "It's nice ridin' with a fellow Georgia boy."

"Yessir, where you from in Georgia?"

"Rome."

"Say what?" I asked, despite hearing his answer.

"Rome, GA," the guardsman replied as he said the letters G and A separately.

"You're a Roman?"

"Huh?"

"Nuttin'." I snickered. "I lived in Rome for a lil while."

"Oh yeah?"

"Yep, worked there for a bit."

"Then you know what God's country looks like." The guardsmen smiled, kicked back in his seat, and propped his feet on ammo cans strapped across from him.

I don't remember the guardsman's name, so I'll refer to him as the Roman.

"Charlie-Six, Charlie-Six, this is Vampire-Six, over." The Georgia National Guard Platoon's lieutenant radioed Captain Mac.

"This is Charlie-Six; go ahead, Vampire-Six."

"Roger, Charlie-Six, I need to inform you that an ANA vehicle struck an IED, break. We have casualties."

"Roger, Vampire-Six . . ." Captain Mac relayed his instructions to the lieutenant. These included dialing up a casualty evacuation and pressing on to Margah under a helicopter escort.

We set up a perimeter at the base of COP Margah and hunkered in for the night. In the AM, we'd clear villages and try to root out the enemy. That evening, while I was talking smack with Brandon and Smitty, a mortar exploded in the distance, and small arms fire sprayed from a nearby hill. Since our vehicle faced enemy fire, Sergeant Bess opened up with the .50 cal. The rest of us took cover behind the MRAP. Mortars detonated into the hill near the COP but not dangerously close. The black sky was illuminated with tracers fired from our machineguns. A screeching explosion went off by the vehicle to our left, and a large puff of smoke curled upwards.

"Holy shit, they've been hit!"

We assumed a mortar impacted by the MRAP. Frankie and Rubio were part of that crew. Following a tense pause, Ric Flair-style "wooos" erupted from the smoky MRAP. The explosion had resulted from the backblast of an M72 LAW (Light Anti-Tank Weapon). It's a portable one-shot anti-armor weapon firing a solid rocket. Rubio ordered Frankie to light 'em the fuck up.

Following the LAW blast, the enemy retreated, and a ceasefire was ordered. I peeked at Bess inside the turret and gave him a fist bump. As Brandon and I replayed the short ambush over a cigarette, we learned an Afghan soldier was wounded during the melee. A bird swooped in to evacuate him. He wasn't hit by enemy fire, though. The round was fired by one of his fellow ANA soldiers. This was why we felt sketchy around these dudes. If the enemy attacks a perimeter at six o'clock, soldiers at or near the six o'clock position should be the only ones returning fire. The guys at twelve o'clock must stand down, or else they'll shoot their own people in a crossfire. The Afghan fired from the twelve o'clock position and struck his own dude at six o'clock. It could've been one of us catching a medevac.

Stan passed out a guard schedule, and we called it a night. I plopped in the dirt beside the vehicle's front tire and dozed off.

It wasn't long before my sleep was interrupted by the Roman calling for Stan from the turret. "Sergeant Stan."

"Yeah, what's up, man?"

"Um, I don't feel comfortable behind this fifty," the Roman said. "I'm not real proficient with it right now."

"Get up there, Smitty."

"You gotta be shittin' me," I whispered to Brandon while lying on my back in the dirt.

"Dude's a fucking sergeant and doesn't know the fifty?" Brandon asked as he sat a few feet away, snacking on an MRE.

"All that shit he was talkin' 'bout training in the guard and his time in Eighty-Second during Cold War," I said to Brandon, and we laughed.

"Yeah, fuck him."

Fuckin' Roman.

A dreary fog blanketed the morning as I rose from the dirt. Once the sun peeked over the mountains, it burned off some of the haze. I mounted and readied for patrol. We were headed deep into hostile territory. The HIDES was stuffed inside my pack again; I'd be dealing with Afghan males. Stan handed me the LAW, too, so I strapped it across the top of my pack and clutched my M4 at the low ready.

Over the next few days, our huge element cleared Margah's battlespace with precision. Standing shoulder to shoulder, we stepped through dense cornfields taller than the tallest soldier. We toured mud homes, one by one, and scanned for suspicious activities, caches of weapons, and IED-making materials. Captain Mac and Manny lined up Afghan men so I could do my thing with the HIDES. The brigade photographer snapped photos as we performed our duties. She never left Captain Mac's side. I assumed these photos would someday make the pages of a school yearbook-type thing that'd polish this turd.

Each day, I ran with Sergeant Rubio. I loved rollin' with Rubio. A no-nonsense type, his skills had been honed through several combat deployments. He took shit from no one, and his attitude towards locals was "distrust and verify." He sniffed out bullshit and shady Afghans like a redbone coonhound. I trusted every move he made.

His methods weren't for everyone, but he got shit done. A homemaker in Cincinnati wouldn't approve. But the way he conducts his war is none of her biz. He refused to treat these people with the kid gloves Uncle Sam issued. Rubio had been on too many deployments and seen too much shit. He understood what politicians, pundits, generals, professors, and academics didn't. They conducted and critiqued this war from their ivory towers while smoking Cuban cigars and drinking fine red wine. We were there on the ground.

In one kalat, I lined men up inside the living room for the HIDES routine. An old, frail man was being difficult. I failed to catch pics of his pupils because he blinked nonstop.

"He keeps blinking, damnit." Rubio strutted by after hearing my frustration.

He grabbed the long-bearded old man by his nappy white hair, yanked his neck backward, and said. "You better open your fucking eyes, motherfucker!"

The man gagged and struggled. "I'll shoot your fucking ass right now, bitch. Fucking try me. If I have to, I'll pull your fucking eyes outta the sockets and scan those bitches that way."

Rubio gazed at me with a hand full of hair and said. "He understands what the fuck I'm saying. All these motherfuckers do!"

I got a solid scan of the old man's pupils during my next attempt

After three days, we saddled up to return to Boris. During the ride back, I couldn't help but reflect on what we did and didn't accomplish. "Y'all knew those cocksuckers weren't gonna fuck with us while we strolled through their hood."

I ranted. "And we weren't gonna find shit, anyway. The enemy knew we were coming, so anything that they had got hidden. Plus, any Taliban fuckers morphed into the poor farmer, artisan, blacksmith, or whatever the fuck these Stone Age people do by day. They shook our hands with a laugh because they knew they'd fuck us later. Hell, if a motherfucker appeared shady and couldn't hide behind a fake smile, y'all know the look, there wasn't shit gonna be done about it. Ole Sergeant Rubio wanted to roundup some shady lookin' motherfuckers. Of course, the brass wouldn't let him."

I was interrupted by a boom. An IED had popped off and struck the lead vehicle. A pickup truck sped off as soon as the bomb detonated. Stan's MRAP was placed behind the point vehicle, and we felt the residual thump while watching the lead vehicle absorb the full impact.

"Fuck, Shane and Frankie are in there," Stan screamed. "Bess, fire that fucking fifty!"

Bess rocked the .50 cal along with Blackie, who was gunning another MRAP. The pickup truck came to an abrupt halt. Two men jumped out, dropped to their knees, and placed their hands behind their heads. Guns ceased firing, and the dismounts fled vehicles. Shane exited the hit MRAP and signaled to everyone that all was good. In Manny Fresh's finest moment, he charged the two Jihadis and boot-stomped one to the back of his head. He grabbed the other man by the neck and planted his face into the dirt.

I stood by and egged Manny on. "Fuck yeah, sergeant. Beat his ass!"

He continued face-smashing the dude. It trumped being ringside at a rasslin' match.

"Do it for 'Merica, sergeant," rang out from the crowd.

When Manny finished, the two shitbags were given to the ANA. They roughed up the dudes before blindfolding and tossing them into the back of a Ford Ranger. Rubio searched the pickup and found bomb-making materials. Captain Mac and the brigade photographer approached the scene, and he offered further instructions while she snapped photos. *Too bad they missed it.*

Two weeks later, Third Platoon rolled back to Margah. We'd relieve Dog Platoon and assist Second with teardown. This time a Route Clearance Platoon accompanied us. It consisted of grunts from a Baker Company platoon, combat engineers, Explosive Ordinance Disposal (EOD) technicians, and all types of funky vehicles I'd never seen. It was good to see the Second Platoon dudes again. We hugged it out and told a few war stories. They wore cool hats with "The Greasers" stenciled on the front, alluding to greasing Hajis.

On the final day of teardown, Second Platoon was flown to Boris via helicopter. Third Platoon and Route Clearance Platoon geared up to convoy back. It couldn't have come at a better time because we were out of water and MREs.

As we convoyed, two Apache attack helicopters swooped in and laid scunion into the side of a mountain named Bushmills. A couple of missiles were launched from the birds and crashed into the mountainside.

"Fuck yeah. Get some. Send 'em to their fucking Allah," rang out.

This was the first time we'd witnessed Apaches reaping destruction on anything. Once fire ceased, the pilot radioed Lieutenant Troutman to inform him they'd destroyed a recoilless rifle and several rounds. It's one of the most feared weapons the enemy possesses. They fire a modified artillery shell yet are lightweight enough to be portable. The weapon can be mounted in the back of a pickup or moved by a small team.

The screams of joy were short-lived, though. The higher command caught wind of the Apache strike, and Charlie Company's executive officer was on the horn.

"Three-Six, Three-Six, this is Charlie-Five, over."

"This is Three-Six. Go ahead, Charlie-Five," Lieutenant Troutman acknowledged.

"Three-Six, we need you to perform a BDA of the Apache strike on Bushmills, break. And we're gonna need you to recover the recoilless rifle, break. Bring it back to Boris."

"Roger, Charlie-Five. But you do realize the recoilless rifle has been destroyed, right?"

"Three-Six, we haven't confirmed it."

"Charlie-Five, give me one moment."

Lieutenant Troutman switched his frequency to communicate with the Apache pilot and asked, "Can you confirm the weapon is destroyed and disabled?"

The pilot did us a solid and performed another flyover of the destroyed rifle. "Three-Six, I can confirm the weapon and all rounds are unusable."

Lieutenant Troutman switched back to company frequency. "Charlie-Five, Charlie-Five, this is Three-Six, over."

"This is Charlie-Five. Go ahead, Three-Six."

"Charlie-Five, I need to inform you that the pilot completed another flyover, break. And he's confirmed the recoilless rifle and all rounds are unusable, over."

"Three-Six, that isn't your concern, break. You have your orders. Do you understand?"

"Roger, Charlie-Five, but I need to inform you the entire element is black on food and water, break. Is the disabled weapon that important?"

"Again, Three-Six, you have your orders, and you need to get started, over."

I sat inside the MRAP, listening to the conversation in disbelief. The Duke was seated beside me, and I said to him. "You gotta be shittin' me, man. They're gonna send us up there and risk us walking into a trap over a goddamn incapacitated weapon?"

"Bullshit, man," The Duke said. "Apparently, recoilless rifles are rare in this AO."

"So, it's a fucking trophy for some commander?"

"Yep, pretty much."

Yes, a commander needed to boost his resume, and he had the power to force us to walk several klicks uphill and risk our lives doing it.

Troutman's handling of the recoilless rifle situation confirmed why I loved him so much. He fought with command and attempted to spare us from unnecessary bullshit. His men were more important than any trophy-gathering nonsense.

At the time, I was not happy with Captain Mac, the executive officer, or even the first sergeant. I blamed them for the denied fire missions, the senseless BDAs, the piss test ambush, and of course, this trophy-gathering bullshit. However, most of these shitty decisions came from above Captain Mac's head. If anything, he did his best behind the scenes. Tons of hate was directed towards the executive officer, too. He was always the bearer of bad news over the radio. We imagined him sitting inside the Tactical Operations

Center (TOC) on Boris behind his desk after eating hot chow. He rarely went "outside wire" and looked like a douche bag when he did.

We prepared to climb Bushmills. But, to my surprise, the Route Clearance Platoon was chosen to perform the task. Lieutenant Troutman and the lieutenant in charge of Route Clearance Platoon agreed they wanted to get this show on the road. The Route Clearance Platoon made it up the hill without drama and did the job asked of them. Their lieutenant brought the EOD guys along and blew up the recoilless rifle in place. He thus assured that the weapon was rendered useless and left nothing to carry down the mountain. I'd love to have been a fly on the wall inside TOC when Troutman and the other lieutenant explained themselves.

Leaving Margah in the hands of the ABP was wishful thinking, though. Our Afghan friends said, "Peace out," after we left the area. Locals swarmed the former COP like ants on a watermelon.

Lieutenant Troutman gathered the platoon to say what nobody wanted to hear. "Guys, get packed and get vehicles ready because we're heading back to Margah. We expect to be out there for three days, so pack accordingly."

We arrived in Margah, and the platoon was divided into two elements. One element pushed to the top of the hill where COP Margah once stood, and the other set up at the half-built ABP compound at the bottom of hill. A wall surrounded the ABP compound, and several unfinished concrete buildings sat inside. A makeshift TOC was created in one building, and the platoon slept on the floor nearby. Two MRAPs were parked on opposite ends of the compound. ABP guys performed guard shifts inside the towers along the walls. The other element of the platoon, atop the hill, positioned two MRAPs about fifty meters from one another facing in opposite directions.

I was positioned atop the hill with Sergeant Rubio and eight others. Frankie, Brandon, Dottie, and I occupied an MRAP. We rotated four hours-on and twelve hours-off guard shifts, sleeping in and around the vehicle between watches. The larger element at the bottom of the hill sometimes pushed out patrols through villages and into the Margah bazaar. We provided overwatch. To my surprise, things remained calm beside the occasional rocket attack. Yet, I felt uneasy.

The third day came and passed. Days rolled by, and we felt like we were being hung out to dry. During the late afternoon each day, we listened to the radio as Lieutenant Troutman called the Charlie Company TOC to ask for a SITREP. Charlie-Five would respond that he had no new information. It was a kick in the balls each day.

One night after hearing the bad news, I blurted out in my finest singing voice. "One more night."

Dottie and Frankie sang along. "One more night," over and over for hours.

Phil Collins' *One More Night* became the motto of the trip. For two weeks, we camped atop Cognac until word came that Dog Platoon would roll out to relieve us in the AM. Two-week rotations in Margah were established between Second, Third, and Dog Platoons. Sure, we were delighted to get back to FOB Boris. But news of continued rotations to Margah didn't sit well.

"Why in the fuck would we tear down a perfectly good COP only to come back days later to man a fucked up brittle compound at the bottom of the hill?" I asked Dottie, frustrated. "Along with a bare piece of dirt where the COP once stood?"

"I dunno," Dottie said between chewing his MRE bread. "They trying to get us killed?"

It appeared command didn't give a rat's ass about us. Why else would they order us to tear down a COP and vacate a volatile area only to return in a diminished capacity? There was no method to the madness of the Margah situation.

Dog arrived, so we mounted vehicles and took off on Volkswagen. I climbed inside the turret behind a .240 Bravo to gun the point vehicle. We were distracted by thoughts of showers, hot chow, the internet, and movies. Brandon taped his mp3 earbud to the microphone of one of the headsets, and we listened to music on the ride. The music stopped whenever someone spoke over the radio, but it returned once the chatter stopped. Alanis Morissette's *Ironic* and Billy Idol's *Rebel Yell* screamed through the speakers, and we all sang along. I sneered my lip and pumped my fist like Billy Idol.

"With the rebel yell, she cried more, more, more. In the midnight hour, babe more . . ."

A thunderous explosion interrupted the music.

The vehicle filled with black smoke after an ear-piercing sound rattled the air. I fell through the turret into the bottom of the seat. My ears rang, and it was difficult to see through the smoke. As I did after the MRAP rollover, I reached to feel my legs and confirm they were still there. They were fine. As I shook cobwebs from my head, the top hatch of the MRAP was flung open, and Stu popped out firing. He inspired me. I reached for .240, which had broken loose from the mount, and swung it to my left. The turret was disabled, so I hung myself outside the edge of the turret, aimed the weapon, and sprayed my own 7.62 automatic fire. Marty popped out of the top hatch and launched grenades through his M203. After burning through a hundred rounds, The Duke handed off another ammo can, so I could reload and continued laying it down. Each time I fired a hundred rounds, The Duke calmly passed another hundred rounds. It wasn't his first rodeo.

While shooting, I could hear Lieutenant Troutman over the radio saying. "Make sure you have a positive ID on targets."

Nothing against my beloved lieutenant, but positive identification was the last thing on my mind. *Fuck 'em, I just got my ass blown up.*

A ceasefire was ordered. With the amount of reloading I'd done, I figured

I'd squeezed off over four-hundred rounds. Troutman, Donald, and Rubio dismounted to inspect the crater where the blast erupted.

"Holy shit, that crater is huge," Donald said.

While the crater was analyzed, Brandon, Frankie, and I polished our story in case the politically correct army reared its ugly head.

"Some dude just took off running after the blast," Brandon said, and Frankie nodded in agreement.

No one from the company asked any questions about the incident. Once the smoke cleared, everyone saddled back up, and the movement continued. Another MRAP pulled around to serve as the point. Blackie was gunning, and I laughed when he glanced at me with a shit-eating grin and flashed the horns. I smiled and gave him the horns right back. Apaches swooped in to provide an escort down Volkswagen. We resumed rocking out.

"Eight months in-country, and this is the first-time helicopters escorted us anywhere," I said.

"Yep, in Iraq, we pretty much had an Apache escort everywhere," Frankie replied.

Following a quick shower, I ventured to the MWR to log on to the internet. Despite feeling concussed, I glowed. Perhaps I was thankful to be out of that Margah shithole? Maybe I felt blessed to be alive? Could it be that I laid some scunion down and was still feeling the rush? How 'bout the thought of watching football tonight? Whatever it was, I was on cloud nine.

I exited the MWR and ran into a buddy I hadn't seen in a while named Clover. He was a good ole boy from Kentucky and had just been promoted to sergeant. He'd served as company armorer, making him well-acquainted with everyone. I was fond of Clover because he was a likable dude who took me under his wing and showed me the ropes when I arrived at the company.

"I heard you had a bad day, Bruce," Clover said.

"Yeah, it got a lil rough."

"Well, I got something that'll make it all good," he said, pulling a bag of hashish from his pocket.

"Fuck yeah, my cracka." Clover and I smoked some fine Afghan Kush.

"Smoking the kush inside the Hindu Kush," Clover said, and we giggled like children.

Clover had his own room away from everything on the FOB. We had no worries. He strummed an acoustic guitar, and we sang classics like *A Country Boy Can Survive*. For a moment, I felt as if I were chillin' at home amongst good friends. The hashish eased my mind.

We ended up getting the munchies, so Clover asked, "Wanna go to the chow hall and grab a bite?"

"Sure, man," I replied. "I wouldn't mind watching some football, either."

Since we were half a world away, college football games played at midnight on Saturday night and continued through the wee hours on Sunday. It reminded me of my time in Okinawa when we watched football past midnight.

Clover and I entered what appeared to be an empty chow hall. We commenced snacking on candy bars set up near the front. I peeked at the television and saw it was on CNN, so I switched it to ESPN. There were two small rooms inside the chow hall, and whenever the channel changed in one room, it switched on the tv in the other room as well.

I heard footsteps from the far room. "Bruce!"

I turned and snapped to parade rest. "Yes, sergeant major."

"I heard you had a rough day, son."

"It's all good, sergeant major."

"You stood back up and returned fire on the bastards. You did good, son."

"Thank you, sergeant major." He patted my shoulder and left.

"Dude, what the fuck was that?" I asked Clover and giggled my ass off.

"I dunno, dude," Clover said. "I was standing at parade rest behind you stoned outta my mind."

"Did you know he was gonna be on Boris?"

"Fuck, I never received any word. He must've flown in unannounced."

"Oh well, we know now," I said, and we both cracked up in stoned laughter.

That Saturday was Halloween. Every Halloween, I think about that bruising IED. Had it exploded any other day, I wouldn't remember the date. It doesn't haunt me, but it's a memory that reappears when I hand out candy or attend parties. Ten years earlier, on Halloween 1999, I was rockin' to Motorhead inside the Masquerade. Halloween to me is not ghouls and goblins, but Motorhead and IED blasts. It's a stoned chat with the command sergeant major that lightens my mood. How metal is that? \m/

CHAPTER 12: NOVEMBER COMING FIRE

I started getting a bad feeling after a QRF mission to assist the special forces guys. We helped them recover an MRAP they'd rolled. Stu and I dismounted to set up a security perimeter while the massive wrecker snatched the MRAP. As Stu and I sat on a large boulder, a couple of "long beards" wandered our way to chitchat. They were more approachable than most NCOs and officers we'd met on Bagram and Sharana.

"Hey guys, how you doin'?" the long beard asked.

"Doin' well, sir," I replied, not knowing the man's rank. "How 'bout you?"

"Oh, it could be better, but we appreciate you guys coming out."

"No problem, sir," Stu said.

I glanced over the "long beard" and recognized his Auburn cap atop his long, dark feathered hair. He reminded me of Randy Owen, the singer of country band "Alabama," in his prime.

"War Eagle," I said before asking. "Did you go to Auburn?"

"I got my master's there," the man said. "War Eagle!"

"They aight," I said and smiled.

The man chuckled and asked. "You from Alabama?"

"Naw, Georgia, but I don't mind Auburn."

"So, you're a Bulldog?"

"Hell naw, I'm a Yellow Jacket."

"Yeah, fuck them Dawgs," the man said and laughed.

These long beards had "Type A" personalities. Most held advanced degrees and attended every specialized school the military offers. The conversation turned more serious when the long beard explained what had been going down inside the AO.

"I hear you guys conduct a lotta BDAs," the long beard said, while clutching his decked-out rifle.

"Yes sir," Stu said. "Pretty much every time we take indirect fire."

"Y'all need to keep your eyes peeled."

"Don't you think doin' a BDA every time we take fire sets a pattern?" I asked.

"Well, that's what I'm getting at." The man stepped forward. "You guys need to be on point during these BDAs because the enemy is starting to set traps."

"Oh yeah?"

"Yep," the long beard cleared his throat and paused for a couple seconds. "They're starting to booby trap POO sites. A guy lost his leg stepping on a mine at a POO, and another was killed during a BDA."

"Fuckin' A," I said. "Hate to hear that, sir."

"Be careful doing BDAs," the man implored before stepping off. "It was good talking to you both, War Eagle!"

After recovering vehicle and hooking it to the wrecker, we convoyed to Firebase Lilley to drop off the long beards' MRAP. When we arrived, the long beards had a special treat – a meal inside their chow hall. Although small in stature, Lilley possessed dope amenities and a fine dining facility. The DFAC employed local chefs and was so clean that I could eat off the floor. The menu was deep and the food tasty.

During the meal, WWE's Monday Night Raw played on the "tube." I caught up with wrestling I hadn't watched in years. The contemporary stars such as John Cena, CM Punk, and Randy Orton were entertaining. But new skool rasslin' paled in comparison to the days of my childhood. I yearned for the days of Tommy "Wildfire" Rich, Mr. Wrestling #2, and the "Nature Boy" Ric Flair during his prime. I daydreamed about when my father took the family to wrestling in the 1980s. ATLiens flooded wrestling's mecca, The Omni, and my father and I rooted for the bad guys as my dear ole mother voiced her displeasure. She'd stand near the rail and scream at a villain as the rugged heel insulted her in return. I reminisced about a trip to the Cobb County Civic Center, as a twenty-two-year-old in 1998, to watch an ECW event. The Dudley Boyz taunted the crowd. Bubba Ray called some gal "a five-dollar blow job whore," and then he told the angry crowd "D-Von fucks white women!" *Ah, the good ole days of rasslin'.*

It wasn't long after this November encounter with the long beards that FOB Boris was again attacked by indirect fire. I was lying in bed listening to Samhain's *Mother of Mercy* on my mp3. I'd been thinking about a concert I'd attended ten years before when Glenn Danzig opened as Samhain—a horror punk band he'd fronted before his solo career. I was rocking the *November-Coming-Fire* album in honor of the month. Stu tapped my shoulder and informed me of the attack. I leaped from the bed and rushed to find my guys. Stan was on leave, making me acting team leader and Stu the acting squad leader.

"BDA time," I predicted to Brandon; it didn't take a rocket scientist to guess.

"Alright, start gettin' it on," Stu said as he reentered room. "Bruce, Blackie, y'all harness up because you two will be gunning."

"Roger that," I said and prepped the .240. Blackie did the same with the Mark 19.

The platoon had a little down time before the mission kicked off, so I fired up a cig and sat by The Duke to chitchat. As I exhaled smoke from my Pines, I peeked inside the room behind me and saw Julian sleeping on a lounge chair. He appeared at peace, even though our patrol was about to roll. I felt uneasy due to what the long beard had said.

"You know these BDA missions are fucked up, man," I said to The Duke.

"Yep."

"You know what one of those SF dudes said when we recovered their vehicle?" I asked.

"What's that?"

After I regurgitated the conversation with the long beard, The Duke said. "No shit."

"I don't feel good about this, man," I said.

"Yeah, I hear you."

Lieutenant Troutman rallied the platoon to offer final instructions and lead us in prayer.

"Gents, we're heading into the Mangritay Valley," the lieutenant said. Troutman reminded us to exercise situational awareness and obtain positive identification of targets. "Dog is in Margah and Second is on a mission near Shkin. We're on our own, gents," he emphasized before requesting: "Let's bow our heads. Dear Heavenly Father . . ."

I posted inside the turret and charged the .240 Bravo machinegun. It felt good being gunner. It was a privilege to stand behind the big gun. I was entrusted to rain hell on whomever I deemed deserving. A grunt must savor these moments behind crew-served weapons. As a grunt advances throughout his infantry career, the big guns get taken away. Stan always reminded us "you guys are the shooters and I direct the fight."

Lines from my favorite gangsta rappers played inside my head each time I gunned. How can they not when you're rollin' through the hood and packin' heat? Sometimes a burning cig dangled from my lips, and I imagined myself as 8Ball, "Ridin' thru the hood with my homies gettin' smoked out," with my finger near the trigger ready to, "shoot a bitch in the dome with my motherfuckin' chrome." If Jihadi stepped wrong, I'd blast his ass.

My MRAP included Manny Fresh as TC, Ross as driver, and Doc Lindley and Frankie rolled in the back. We were fourth and last in the convoy, so I faced backward and scanned the rear. Four pickups filled with ANA guys piggybacked behind us. We approached the Mangritay Valley and prepared to enter the village of Sharqi Mangritay. I noticed that the land seemed to swallow the small wadi in the middle. The hillside was steep on both sides of the wadi. I peeked each direction and all I made out were hillsides. The one to my left was dotted with trees, and the opposite hill housed rows of kalats. The convoy moved through the low ground. We had none of the advantages an infantry leader would desire. There were several narrow chokepoints, which hindered our combat power. The enemy had a superior tactical advantage with the high ground and the element of surprise.

The valley cracked with a vicious explosion while we were knee-deep inside the narrow wadi. Deafening thunder roared. The point vehicle, led by The Duke, triggered an IED, and a ferocious barrage of enemy fire followed. Mortars, RPGs, small arms, and automatic machinegun fire rained down. The Jihadi had lured us in. The eruption of the IED signaled the start of the enemy's ambush. Simultaneously, an RPG impacted the driver's side door of the second MRAP. The hull of Lieutenant Troutman's vehicle had been penetrated; we'd been sucker-punched and bloodied by a devastating one-two combination. Yet we weren't knocked out yet. The initial onslaught was grim. The enemy held high ground and surrounded us on three sides.

Following the initial explosions, I clutched the pistol grip of my .240, peered through its optics, and squeezed the trigger. I "searched and traversed" hillsides while a heavy volume of 7.62 automatic gunfire blasted from my hands.

Frankie popped his head outside the top hatch to survey the damage. "Dude, I'm sorry," I screamed as I unknowingly swung the weapon by his head and squeezed off a volley of rounds.

"It's okay, dude," Frankie said. "Do your thang, baby!"

Under a cloud of bullets, Manny, Frankie, and Doc Lindley dismounted the MRAP. Some guys rallied to maneuver toward the enemy while others sprinted to vehicles to aid the wounded.

"It's just you and me now, Ronny B," Ross screamed into the turret from his driver's seat.

"Let's get 'em."

Ross and I fought on alone but in solidarity with each other. Throughout the firefight, Ross positioned himself outside the driver's side door with his weapon propped on the roof. He scanned the area through the powerful optic on his M14 and called out areas for me to shoot.

"Aim to the two o'clock, Ronny B!" Ross screamed. "Walk it down a lil more, buddy. There you go. Move it to the one o'clock and be careful of our vehicles ahead. There you go; put some rounds right there. Now swing it around to the ten and light up those fucking houses. Lookout for Blackie and Brandon."

Ross also communicated with the company TOC. He provided updates and answered questions from Captain Mac because Manny Fresh was nowhere to be found. Manny's broken English, coupled with the chaos of the situation, made it impossible to understand his radio chatter, anyway.

As I fired away, I observed Blackie rocking his Mark 19 at the kalats to his left. Explosions from the grenades peppered homes. I looked on as Brandon, the MRAP driver, climbed onto the roof and stood tall, laying scunion through the barrel of a SAW. After Brandon burned through SAW rounds, he snatched Blackie's M203 and pumped grenades into kalats. Exposed and under no kind of cover at all, Brandon stood atop that MRAP and fired every available weapon at his disposal.

Mortars whistled our way and exploded midair. It was obvious these air burst mortars were designed to takeout machineguns. The Jihadis were all in, and we were still surrounded. If they could close, they'd attempt to kill every one of us and drag our carcasses through their filthy villages.

Elsewhere, Phil worked his .240 from the roof of an MRAP like a pro. Stu paced and tried to get the ANA into the fight. They'd dismounted their pickups and hid inside a nearby ditch.

"Get the fuck up and fire, you assholes!" Stu screamed at the Afghan cowards.

When a gunner dropped due to his wounds, Stu climbed into his turret and ensured that heavy machinegun fire continued. Trubisky communicated with pilots as he frantically attempted to get birds to us. Doc Lindley treated the wounded and Frankie moved about on foot, attempting to gain a tactical edge. Shane, the Pennsylvania grappler, communicated with the TOC and dialed up a casualty evacuation to remove our wounded.

"Goddamnit, Ross, that was close," I said as a round zipped by my head and impacted inside the turret.

Pings from bullets hitting our vehicle rang inside my ears. Rounds struck inside and around the turret where I stood. "Fuck!" I screamed as another one hit inside the turret and ricocheted around.

I didn't realize until later that the bullet ripped inside the back of my armored vest.

I kept firing a hundred rounds at a time, reloading, and firing again. Ross continued to guide my fire. When I spotted people moving in ridgelines and around kalats, I fixed my barrel and unloaded. If the initial burst missed or didn't drop 'em, I squeezed, searched, and traversed, until the Jihadi hit dirt. Then I double-tapped his ass on the ground with another burst and moved on to the next target. Adrenaline flooded my body as I mowed 'em down. But the heavy volume of fire overheated the .240 barrel and caused the machinegun to malfunction. I went to work, locked and cleared the weapon of a failure to extract. As he were sent by God, Shane appeared and tossed some lubrication into the turret. After a lull, Daddy was back in the saddle. More insurgents emerged, and more got put to sleep.

"Shoot 'em down, shoot 'em down. Shoot 'em down to the ground," played inside my head, as Lemmy's cigarette-fueled voice growled this Motorhead cover of a Twisted Sister tune.

Following a bloody exchange, we gained fire superiority. The enemy fire slowed and ceased. We'd knocked out the insurgents' indirect fire assets and prevented the Taliban's elements from surrounding us. Our jets swarmed the area, and the insurgents retreated. The mere presence of airpower was enough to intimidate most insurgents. The firefight's ending couldn't have come at a better time, too. My .240 was out of ammo. I'd fired the fifteen-hundred rounds stored inside the vehicle. All I had left was my M4 and one-hundred 5.56s.

I witnessed someone pulling a body from the second MRAP and loading it into another vehicle. Sergeant Donald had been handling the wounded, and his uniform and armor was covered with an insane amount of blood. He even jumped in and drove the hit vehicle after the driver had been critically wounded.

I heard Manny say over the radio. "Weez gozza KIA." (Killed in Action).

My heart sunk to the pit of my stomach. But there was no time to process the horrifying news. Still "outside the wire," we needed to get our wounded guys to the hospital. We were ordered to leave the Mangritay Valley for a safer place to evacuate casualties. I stood inside turret with the M4 in tow. As we departed, I squeezed off a few rounds, drive-by style, into the kalats to my front. Glass shattered as I nailed windows.

Manny Fresh said, "Ceez fire Vruce." He appeared to smirk.

Shane confirmed a casualty evacuation, and we set up a perimeter outside the valley. Everyone dismounted to assist with casualties. Second Platoon also arrived on scene. They'd mounted up and hauled ass from Shkin once they learned we were in trouble. I watched D-Rum and Trubisky limp to the bird with minor wounds. Then I noticed a shirtless Troutman being assisted to the chopper. His shoulders were covered with a giant bandage. Our beloved lieutenant was being whisked away from the battlefield. Two other men carried a stretcher with a body covered with a poncho to hide the face, our KIA.

I looked at Ross. "Who's the KIA?"

Following a sigh and a deep breath, Ross said in a choked-up voice., "Julian."

This will sound cliché, but Julian was one of the good guys. He possessed an unwavering faith in God and was a devoted family man. What makes it worse is that Julian was scheduled for the next R&R rotation. In fact, the R&R bird arrived at FOB Boris the morning after the battle. Julian's daughter Mya would be celebrating her first birthday. Instead of arriving home to celebrate, Julian would be flown to West Virginia inside a flag-draped coffin.

Before deploying, Julian emphasized that he wanted to fight for families like his. He'd made his impact on this earth and touched the lives of all who knew him. I aspire to be a man of his caliber. Often, I wished I could take his place.

I was the old man in platoon. I'd lived more years. I had no children nor family of my own. Julian had all that.

As the firefight went down, Stan and Hix were at Bagram returning from leave. An NCO from the 509th informed them that something was going down with Charlie Company. They sprinted to the Bagram TOC and listened to the company frequency as the battle was being fought. Casualty information came through and Stan read codes off the battle roster to see who was being evacuated. A KIA was announced, and Stan reached for his manifest cards to identify who was killed. He and Hix ascertained it was Julian. Anger and guilt overwhelmed them, and they broke down. Their sadness was coupled with resentment because they stood with fobbit officers rather than side-by-side with their brothers in arms.

The medevac bird flew into Bagram with Lieutenant Troutman onboard. Stan and Hix were there to greet it. They approached the wounded lieutenant to offer moral support. Following a brief meeting, Lieutenant Troutman was rushed to emergency surgery. Stan and Hix waited inside the hospital until they heard Troutman's condition. After a few hours, they were allowed to visit. Stan and Hix attempted to cheer up the injured lieutenant. He was in bad shape but would eventually pull through.

Command informed Stan and Hix that Julian would be flown home the next morning. The Third Platoon guys who were present would load Julian's casket inside the bird. Joey C and Wimes were also returning from leave, so Stan gathered them to assist.

An officer from the 509th pointed out a soldier and asked, "What about this guy? Wasn't he in your platoon?"

The soldier was Phife. One morning, Hix strolled to the MWR, and, to his dismay, he spotted Phife munching on Mickey D's and playing on a computer.

Hix slammed the computer shut and yelled. "Get the fuck up, bitch!"

Shaking with fear, Phife couldn't speak a word. Hix dragged him to the nearest MP (Military Police) to be detained. After months of being AWOL, Phife had decided he would head back to Afghanistan to either rejoin a platoon that didn't want him or accept his punishment. Yet, Phife had suffered cold feet and hung around Kuwait over a month—in transition and unaccounted for. There's no telling how long Phife could've remained in limbo had Hix not located him. The battalion commander ordered Hix to escort Phife to Bagram.

When the officer asked if Phife could assist with loading Julian on the plane, Stan said, "If that bitch gets anywhere near that casket, I'll beat him so badly he'll never walk again."

The next morning Stan, Hix, Joey C, and Wimes were driven to the mortuary to retrieve Julian's body. They loaded him into the back of an open top Humvee and sat on benches surrounding the casket. As they were driven down the runway, everyone on Bagram stood by and paid respects. Generals, admirals, colonels, commandos, they stood in silence. Hix, Stan, Joey C, and Wimes were allowed to pay their respects one final time as the flag-draped coffin was loaded inside the plane.

With tears streaming down Stan's face, he knelt beside Julian and bowed his head. "I'm sorry, Julian. I'm sorry I couldn't be there for you." Stan had one promise to make yet had no idea if he could make good on it. "I'll find those motherfuckers and make them pay."

Following the battle, we returned to Boris filled with grief. Some of us headed to the chow hall. There was nothing better to do. We entered dirty, beaten, battered, and bruised. My patrol shirt was ripped down one arm from my armpit to my wrist. After the firefight, I had sliced it with a knife thinking a ricocheting bullet had grazed my arm. Julian's best friend Marty entered the room. I wanted to say something wise and comforting, but I couldn't, so I clutched him in a hug and said sorry.

Rusty and I sat by one another and made small talk. I focused on the framed picture of Afghan President Hamid Karzai that hung from the wall. "I'm tired of looking at that piece of shit."

Rusty walked to the picture and turned it around, so Karzai's image faced the wall. The next morning, Rusty and I strolled inside the chow hall for breakfast and, lo and behold, that motherfucker's picture was visible again. I voiced my displeasure.

That night as I sat with Blackie, T-Love, and Brandon smoking a cig and reminiscing over Julian, Rusty appeared in the darkness. "Hey Bruce, come here, man."

He led me to his room and had a huge surprise. "My cracka, you got that fucking Karzai picture!"

We carried it to the latrine, propped it against a Hesco wall, and destroyed the picture once and for all. I lifted a massive boulder, slammed it onto the frame, and smirked as the glass shattered. Rusty grabbed a handful of smaller rocks and hurled them. Ole Rusty pulled out his cock and pissed on it all as a finishing touch. Everything was left in place so the local trash man could deal with it during his morning run.

About a week after the tragic firefight, a memorial ceremony was set up on Boris to honor Julian's life. A service was held inside a courtyard under the open sky. To the front of our makeshift chapel sat the classic fallen soldier's monument – a pair of boots at the base, a rifle facing barrel down with Julian's dog tags hung from the pistol grip, and a helmet sitting atop the rifle's buttstock. Julian's name was etched into the helmet's band. A framed picture of Julian sat at the base alongside two medals: a Bronze Star with a Valor Device and a Purple Heart. Flanking the fallen soldier's monument to the rear was an American Flag and a unit flag. "Lest We Forget" was painted along the base.

We dressed in our cleanest uniforms to say goodbye to our fallen brother. It was a clear day with blue skies. Two sets of pews were lined up, and a pamphlet was passed out as we entered. It listed a schedule for service, speakers, a short biography, and the Bible passage Psalm Twenty-Three.

Rusty and I shuffled into the pews reserved for Julian's platoon. Several birds flew in to drop off VIPs, brigade and battalion commanders, sergeant majors, and important civilians. VIPs were seated opposite our platoon, and I noticed they were accompanied by personal bodyguards.

Afghan commanders and troops from the ANA, ASG, and ABP were also in attendance. They paid their respects to a soldier who sacrificed for their country. There were more people on FOB Boris that day than I'd ever seen. It was the first time I saw the brigade commander set foot inside the Bermel Valley.

A beautiful service played out. The Chaplain preached a fine sermon that eased my mind, and Captain Mac choked up as he spoke about what Julian meant to Charlie Company. Julian was eulogized by his squad leader Staff Sergeant Donald, team leader Sergeant Fink, and his best friend Marty. We laughed at the joyous memories and cried at the thought of our brother being gone forever. Some of us were at ease, feeling Julian was with the Lord.

A final rollcall and playing of "Taps" concluded the service. The first sergeant shouted. "Sergeant Adams."

"Here, first sergeant!"

"Specialist Bass,"

"Here, first sergeant!"

"Specialist Berisford. Specialist Berisford. Specialist Julian Lee Guthrie Berisford!"

"It is my duty to inform you that Specialist Julian Lee Guthrie Berisford was killed in action on November 4, 2009, first sergeant!" Sergeant Rubio screamed as part of ceremony.

Afterwards, the bugler went into his rendition of "Taps." The chaplain dismissed everyone, and we were given a chance to pay our final respects. Of course, VIPs were first. They soon headed to the birds.

Rusty whispered. "Did you see the fucking colonel keep checking his watch as if he had better shit to do?"

"Yeah, fuck him."

"A dead soldier is just a number." Rusty said. "Fuck him."

"Yeah, fuck all the VIPs."

Two at a time, Charlie Company dudes stepped to Julian's memorial. We gave a slow hand salute, ripped a name tape from our blouse, and placed it at the base of the monument. I witnessed guys kneel and bow their heads in prayer. Other men gripped Julian's dog tags and murmured a final message. Since there were people of all faiths and no faith, everyone had their own personal method. That was the beauty of the ceremony. All the artifacts left at the monument would be sent home to his family. I saw his "Taliban Hillfighter" patch he loved to wear, a West Virginia Mountaineer hacky sack, bracelets, coins, and notes. It was a fitting display by brothers to honor his short life. Cheers to you, Julian.

CHAPTER 13: THIRD PLATOON GOES ROLLING ALONG

In the early morning of November 5, 2009, Manny Fresh and Smitty took off on leave. Troutman, D-Rum, and Trubisky were being treated in a hospital. Julian was KIA. Jordy, KG, Nate, and Smeed were recovering elsewhere and no longer in-country. Phife, well fuck him. It'd be at least a week before Hix, Stan, Joey C, and Wimes could leave Bagram. They were told the air was red, meaning nothing was flying due to antiaircraft artillery. Third Platoon was running on fumes.

Second Platoon had been inside the Mangritay Valley patrolling since the firefight ended. We were alerted for a possible QRF because Second was having problems with a vehicle. Everyone rushed to prep the MRAPs for another mission. Tons of spent brass and links littered the MRAP I'd gunned. Sergeant Donald approached with mission's manifest; The Duke and I were slotted for guard duty on Boris rather than QRF. We weren't complaining.

That night, I sat inside the tower and death-gripped a small radio. I was listening for updates and hoping there'd be no drama in the valley. There wasn't; the downed MRAP was hooked up and towed back to Boris.

Stan and Hix soon returned with Joey C and Wimes. Leadership voids were being filled, but we were still missing a platoon sergeant and lieutenant. While catching up, I slipped on my armor to show Stan and Hix my near miss.

After a laugh, Stu said. "Look at this shit, sergeant." He pointed out the tears where a bullet ripped through.

"You're a lucky motherfucker, Bruce."

Stu was the one who noticed the torn cloth after I'd suited up for guard duty. I had no idea. Rusty picked out a piece of the bullet fragment lodged inside the turret when he prepped for another mission. He wanted me to have it, so he placed the piece of metal onto a pack of cigs by my bed. When I entered my room later, I snatched the pack of cigarettes to grab a smoke and lost the tiny fragment.

Not long after Stan's return, he pulled me to the side to hold a serious conversation. "I wanna let you know that I'm gonna hold you back from missions and do my best to keep you inside the wire. You deserve to go on leave, man."

I was scheduled for the last R&R rotation, which was next month. Stan and the company leadership didn't want what happened to Julian to happen to me, especially this late in the game. After nine straight months in-country, I was ready to relax.

I was appreciative, but I would've been remiss had I not offered. "Sergeant, if y'all need me, I'm here."

"I know, man."

"Thanks."

"You deserve it."

The truth was that I was physically and mentally drained. For nine months, I'd been living the suck, hiking mountains, embarking on missions, witnessing death, and dodging rockets, mortars, RPGs, small arms, and IEDs. I enjoyed the bonding and brotherhood. Yet I was on edge. My body ached and the grind was catching up to me. The rollover accident, combined with over a hundred foot-patrols, did a number on my back, knees, feet, and joints. My exterior looked like a million bucks, but inside I was wrecked.

I also grew disillusioned with our efforts. It was exhausting to perform missions with both hands tied behind our backs. The counterinsurgency, "winning hearts and minds" game, was a loser. It might look good on paper but it's restrictive to grunts. The tiresome BDAs, being lured into traps, was getting old and deadly. In infantry school, it was all about advantages—outnumbering the enemy, displaying overwhelming firepower, avoiding disadvantageous situations, and maintaining a tactical edge. We did little of that.

Ole Ronny B needed a break. I needed a little Hotlanta. I needed a drink. I needed to see my father and sister. I needed the company of a sweet young thang.

The added downtime lent itself to deep thought. Since I'd been pounding dirt for some time, I felt I had an insight that few possessed. I observed locals, broke down customs and courtesies, learned demographics, studied politics, and fought battles. No media outlet or Washington politician could gain an infantryman's expertise regarding Afghanistan. But no one gives a fuck what the grunt thinks.

During a conversation with Joey C, Rusty, D-Rum, and others, I imparted my wisdom to my young buddies. "Y'all know this whole thing is fucked up, don't you?"

"Whatchu mean, Ronny B?" Joey C puffed his cig as he slumped on the wooden bench in our nighttime gathering spot.

I proceeded with my rant. "This whole fucking thing, man. What are we doin' piddling around here with these fucked up ROEs, gettin' picked apart, and gettin' people hurt and killed? For what? For this shithole? There's no way in hell one of these piss-ant Afghans we're fighting could make it across the pond to do harm in America. You think any of these fucking Hajis, living in their mud homes, are any threat to our friends at home? Man, these people live in the Stone Age. I can understand why we invaded after 9/11, but to still be here over eight years later? For what? This is a nation-building project, my crackaz, and nation-building is bullshit."

"You're right, Ronny B," Rusty interrupted as he stood with his arms crossed. "We're not hunting terrorists; we're passing out soccer balls. And getting shot at the whole time."

"Right on, my cracka."

I continued. "You know how after World War II we assisted in rebuilding Japan and Germany into modern countries with global economies? Do any of y'all think Afghanistan will ever be rebuilt into another Japan or Germany? C'mon, man. The literacy rate here is like twenty percent. Those motherfucking ANAs we've been working with? Their literacy rate is around ten percent. They can't even read fucking maps! Grown men fuck little boys with no shame in it, women are considered lower than dirt, and marriages are still arranged between grown men and child brides. What the fuck are we gonna do with that?"

"Then what are we doing?" D-Rum asked. He toked on a bench beside Joey C.

The young men stared at me as they awaited my answer. It was as if I held the keys that unlocked all secrets.

I thought for a second before replying, "Ah, I hate to disappoint boys, but Daddy don't know."

I thought again. *Should I lead these guys down this rabbit hole? We still have work to do. I've already gone this deep; might as well not leave 'em hangin'.*

"It's complicated," I said. "There isn't one thing that explains it all. Besides, it doesn't matter, anyway, because we're still gonna be here. But I'll let y'all know what I think. I could be wrong, though. I ain't Mr. Know It All."

I took a drag off my cig and continued. "It's all about occupying, my brothers. We're not here to win battles or to win the war; we're here to occupy. There's something here the shithead politicians want, and they're not gonna get it if we pack our shit and go home. I'm not gung-ho by any means, but don't y'all think we'd be a lot safer now had we started at one end of this AO and kicked in every door, cleared every home, and put the fear of God in these people? But no, we roll around waiting to get shot at."

"So, what's here? Well, there's precious minerals and metals to be mined. Some people believe the reserves could be worth trillions of dollars. Trillions of dollars, my crackaz! There's also plans to build a Trans-Afghan Pipeline that'd transport natural gas right through this country. Oh, and by the way, ole Karzai was a top adviser to the corporation planning the pipeline. Ain't that lovely? Don't y'all also realize Afghanistan produces ninety percent of the world's opium? Ninety percent of the world's opium, fellas, and biz has been good since we've been here. We're standing in the illicit drug capital of the world. Colombian and Mexican drug cartels are child's play compared to these dudes."

"Y'all have seen this fucking shithole up close and personal. Does this godforsaken hellhole look like the kind of place that holds the keys to American freedom and democracy? A place that happens to be the place where ninety percent of world's opium is produced? Where trillions of dollars in mineral reserves sit underground? If y'all think we don't have our hands on a lot of that cheddar, you're crazy. The name of the game is to occupy, fellas, and get what they need. We're not here to save the Afghans and defend American freedom. That's horseshit."

I couldn't bitch without offering a solution, though. I didn't want to leave the boys hangin' with doom and gloom.

"There are terrorist groups who've camped here like al-Qaeda. Yes, they hate us, and the Taliban harbors them. The Taliban we've been fighting are no threat to Americans at home; they just wanna run this country. But some groups they're harboring would love to get across the pond and fuck more shit up. If it were up to me, I'd leave a small number of Type A types inside this country to ensure 'The Stan' doesn't train and harbor more terrorists. That's it. The CIA, OGAs (Other Government Agencies), black ops dudes, and the long beards we've met are sufficient. Having over a hundred thousand troops involved in this project is ridiculous. During the first Gulf War, we defeated a large Iraqi Army quickly. Yet over here, we're bogged down by insurgents. A couple guerillas with RPGs and IEDs can inflict a lotta damage, especially with our shitty ROEs. Just leave the Type A's here to fight the guerilla war and prevent this place from existing as a terrorist haven. Fuck the nation-building and all the other bullshit."

I had a lot of time on my hands at this point. I read articles about the country and the war. But most of my opinions derived from my experiences pounding the dirt. I had chosen to be here, so no one needed to feel sorry for me. Sure, I bought into the bullshit from politicians about why we fight. I gobbled it hook, line, and sinker. But it took putting my feet on the ground to understand the gravity of the situation.

I was in the MWR reading articles when I came across a letter written by former marine corps commandant, General Charles C. Krulak. It was a response to conservative columnist George Will's call for the withdrawal of US forces from Afghanistan. My eyes grew wider. General Krulak served as commandant during my marine corps enlistment. He'd swoop in via helicopter before speaking to a giant audience on Camp Pendleton. I listened as he told tales of tours of duty in Vietnam and being wounded twice by the

Vietcong as a young captain. Krulak wasn't large in stature, but he was tough as fuck. Enlisted guys loved him because he rubbed elbows with junior marines. I almost fell out of my chair when I read that Krulak called the war effort a waste of time. He blasted the idea of nation-building and said the war was bogging down troops and equipment without having a clearly defined enemy. Krulak's solution was placing "hunter-killer" teams along the Pakistan border rather than maintaining a large troop presence.

Well, no shit. I felt validated. This wasn't some liberal antiwar hippie; it was General Krulak. My former commandant was saying to get the fuck out of Afghanistan now. *Thank God, I'm not the only one who sees the truth.*

Days after Julian's memorial service, FOB Boris received incoming rockets. The usual artillery counterfire commenced, and the platoon grew anxious. We knew what time it was. Captain Mac ordered Second Platoon to conduct a BDA while Third served as QRF. Second took a tactical approach, driving into the Mangritay Valley on a bone-chilling cold night.

As Second Platoon moved toward the objective, Apaches flew overhead. They observed what was believed to be Taliban fighters setting up an ambush. The pilots couldn't positively identify the weapons, so Captain Mac decided to head out and assess the situation himself. The squad leaders arrived at the cans to spin up the boys. I assisted with mounting machineguns and readying vehicles since I wouldn't be going out. Stan requested to TC the point vehicle in an effort to lead from the front and dispel any fears. He was eager to exact revenge on Julian's killers. Captain Mac rolled inside the back of Stan's MRAP with his entourage in tow. It's rare for a commander to position himself on point. With Wimes driving and Marty gunning, the convoy crept into the darkness.

Wimes trembled and shook. His breathing was heavy, and he stuttered as he spoke into radio.

"Relax," Stan said. "Everything's gonna be fine."

Third Platoon reached the outskirts of the village at one in the morning, and news came in that a dismounted Second Platoon element was having difficulty. The temperatures had dropped, the wind howled, and a couple dudes from Second Platoon had hypothermia. They didn't expect to be out at night and failed to bring extra cold weather gear. Their lieutenant radioed Captain Mac and explained his element was combat ineffective. Guys needed to get inside trucks to warmup. In response, Captain Mac pushed the lieutenant hard to stay in position. But the young lieutenant stood his ground. The two bickered back and forth until Captain Mac relieved the lieutenant of his command over the radio. *You're fired.*

Eventually, the Apaches identified the weapons and dropped ordinance on the insurgents gathered in the hills. Later, a fixed wing bird dropped a thunderous five-hundred-pound bomb that lit up the mountain like a Christmas tree. I could hear the boom and see the flash from the guard tower. To add insult to injury, the Apaches laid down fire as remaining Taliban fighters were mowed down. The carnage ended at about three in the morning, and Captain Mac decided to let the mountain cool. Third Platoon would patrol the bombed-out area once the sun rose.

Stan led the platoon into the hills to survey the area. Slow and tactical, Stan, Hix, Frankie, Fink, Donald, and Rubio swept through and came across the first body—a crispy critter shredded to pieces. The holes in his body were visible from several feet away. As the patrol continued, Rubio grunted in anger. He wanted to be the first to crest the hill because he knew there'd be Hajis ahead to shoot. Further uphill, Hix happened upon a body he'd recognized. On an earlier walkabout through the bazaar, we'd walked into a restaurant full of shady motherfuckers. Hix had begged command to let us round the dudes up and interrogate them. But his request was denied. Oh well, one of those shady motherfuckers was mangled and tossed into a tree. The ANA searched each deceased body and hood-winked the opium.

During one search of the bodies, some ANA dudes got worked up. They found a live one who was thought to be holding a grenade. Stan and Frankie rushed to look. The stranger's back was to Stan and Frankie, so they drew weapons and begged the man to show his hands. He refused. Neither Frankie nor Stan wanted to cap the man in his back. Instead, Frankie picked up a rock and fired away with his arm that'd once played junior college baseball. He missed. Pissed because he'd missed, Frankie grabbed a bigger rock. He tossed a fireball that crashed into the back of the man's head and sent him violently to the ground. As the Afghan lay crying, he was searched and taken into custody.

Stan had knelt to take a break when Harlen approached to tell him that he'd misplaced the HIDES. Tired and pissed, Stan, Hix, Rubio, and Frankie walked outside the perimeter to search. Out of nowhere, an insurgent crept upwards from the prone position. Stan screamed for the insurgent to show his hands, and weapons were drawn. The guy reached for his RPG. Frankie fired two shots. One missed, while another struck the dude yet failed to put him down. With reckless abandon, the insurgent fired his RPG. The round flew overhead, and Stan felt the "whoosh" from the flying rocket. Frankie and Stan opened up with a ferocious barrage. They emptied their mags into the Jihadi and smoked his ass into the ground. Several feet behind the Jihadi, another insurgent laid down, incapacitated from the backblast of the RPG. Ole Rubio smiled, strutted towards backblast dude, and fired two into his face to finish him off. Captain Mac ordered Third Platoon to roundup all the dead bodies and carry them to the ANA vehicles to be taken to Boris. Before Stan lifted the dead Haji that he and Frankie put to sleep, he slid the man's cheap watch off his wrist and placed it inside his pocket.

Everyone returned from the patrol stoked. I stormed out of my room to get the scoop.

I approached Stan and Frankie, grinning. "Tell me about that Haji y'all smoked."

They laughed and described the events with modesty.

"Fuckin' A, man."

Hix opened the door to his room and blasted some loud and fast heavy metal. Everyone was jacked despite being awake for over twenty-four hours.

"About fifty of those motherfuckers met their goddamn Allah last night!" Hix screamed.

I glanced at Hix, flashed the horns, and said, "Fuck 'em."

After getting picked apart all this time, we finally got ours. Stan was reeling from the events; he thought about how his "stop-lossed" ass would be leaving the army soon. It was the eighteenth of November, and he'd turn twenty-five the next day. Never again would ole Iron Guts feel more alive than he did standing atop that mountain. His promise to Julian had been delivered. *We got 'em.*

CHAPTER 14: WHEN RONNY COMES MARCHING HOME

When Johnny comes marching home again
Hurrah! Hurrah!
We'll give him a hearty welcome then
Hurrah! Hurrah!
The men will cheer, and the boys will shout
The ladies they will all turn out
And we'll all feel gay
When Johnny comes marching home

-Patrick Gilmore (1863)

I hadn't been home in over a year, so Georgia was on my mind. When guys spent two weeks at home on R&R, they were out-of-theatre for about a month. The extra time was spent bouncing around FOB Sharana, Bagram Airfield, and some sweet camp in Kuwait. During Vietnam, Big Ronny B spent two weeks in Malaysia for his R&R. Back then, they were sent on vacation but not home. Going from wartime to normal, everyday America would be interesting.

I knew a lot of people would want to hangout. Social media had connected me to past friends and relayed my status. Old friends knew I was deployed. Social media also opened my eyes to how people felt about the war. Young folks in general didn't have the guts to step in your grill, spit in your face, or call you a baby killer. Instead, they said those things on social media behind keyboards and fake profiles. Their "fuck the troops pages" were popular.

Posts included: murderer, slaughterers of innocent people, robots, dumbasses who couldn't get into college, PTSD basket cases, and most were written under the banner of—#fuckthetroops.

Some folks acted hostile towards service members to vent their dissatisfaction with the war effort. Instead of blaming the government and its policies, they rush straight for the throat of a person in uniform. But it's silly to compare the modern soldier's homecoming to that of Big Ronny B and his Vietnam brethren. They returned to a hostile nation, where they were spit on and shunned from society. I could deal with a little flak on social media.

When my friends asked to see me through social media, I knew they would want to hear stories. I thought over what may be appropriate. I knew I'd confide in Big Ronny B and my closest friends. But what would I say to a group of old acquaintances gathered in a bar? Would I tell them what they wanted to hear? Collecting trophies, deserters, the piss test ambush, COP teardowns, and operating under shitty ROEs were a buzzkill. I bet they thought we went on noble terrorist hunting missions – kicking in doors and rounding up bad guys before their plans to attack came to fruition. *I hate to disappoint.*

One thing I did know was my body would be lean and ripped. In addition to grueling missions and intense workouts, I'd been taking a pharmaceutical grade anabolic called Stanozolol. It was used illegally by athletes to gain an edge. Online European pharmacies were happy to ship these products straight to the FOB. I wasn't hardcore enough to purchase injectables, and I didn't need mass building supplements, so I took a boost from an oral tablet to get shredded. I'd never taken an illegal body building substance before Afghanistan. But the ripe age of thirty-three seemed like the time to experiment, so I timed the six-week cycle to end days before boarding the freedom bird. I wasn't the only one receiving a boost, either. Gyms on COPs and FOBs were filled with dudes crushing beast-mode workouts with the aid of needles and pills. I wouldn't call it an epidemic, but it was a widespread dirty little secret. The army didn't give a fuck since our physical performance improved. Some dudes went overboard with the juice, though. A 509th admin sergeant I hadn't seen since Alaska was jacked as fuck when I spotted him on Sharana. His gains were freakish, impossible to obtain in less than a year.

When I noticed him, I smiled and asked. "The fuck you been eating, sergeant?"

"It's the eggs over here," he winked.

Phil, Fink, and I were scooped up with a few others in the company. After a few days in limbo, we made our way to Bagram. It was a massive wonderland that included amenities like Burger King, Green Bean Coffee, Pizza Hut, a massive PX, massage parlors, and pimp-ass chow halls. Signs stated – Country Line Dancing Night Thursday Night at the MWR or Meet Tito Ortiz at the Pat Tillman Memorial USO.

KBR (Kellogg, Brown, and Root), a logistics company, dominated Bagram. They ran dining facilities, MWRs, bus lines, new construction, electrical systems, water facilities, and security. Knowing KBR was a former subsidiary of Halliburton, where former Vice President Dick Cheney served as CEO, fanned the flames of my animosity.

"They're making a ton of cheddar here," I said to Phil. "And ain't even trying to hide it."

"Motherfuckers are getting' paid."

As Phil and I stood in the smoking area in Kuwait, a gentleman stopped and asked. "Do you guys know anything about that DUSTWUN kid?"

The man recognized our patch and understood we were in the same unit as Bergdahl. He wore a civilian tracksuit, and his hair was long and dark. He sported the typical operator beard.

Phil said. "Nope, all we know is he walked off his post."

"Well, I'm looking for him," the man said, and he disappeared in the desert sand.

One week and a few stops later, we boarded the freedom bird in Kuwait. There were two routes, one to Atlanta and one to Dallas. Everyone would disperse to their hometowns from those two airports. After the flight landed in the ATL, Phil and I hit the upward escalator. At the top of the escalator, there was a roped off area with people standing behind a barrier. They held "welcome home" signs and cheered.

A lieutenant general appeared, shook our hands, and said. "Good job, soldiers. Welcome home."

My father and sister rushed over and clutched me in their arms. Phil shook my father's hand, we exchanged goodbyes, and he went to catch a flight to Detroit. All I wanted to do was slip out of uniform because I didn't want more attention drawn to myself, but I knew my father would stop by the Dwarf House in Hapeville, south of the airport. It's the original Chick-fil-A. It'd been a long-standing tradition dating back to my marine corps days.

At my father's home, dressing in civilian clothes was glorious. Beer tasted magnificent. My father's girlfriend at the time had a son named Logan, who'd recently served in the infantry in Iraq. Big Ronny B had taken him under his wing. He helped the young man get reacquainted with regular life and got him hooked up with the VA. Logan and I sat outside on the patio and spoke about our experiences over a few beers. In the evening, Big Ronny B drove us to a local watering hole on Atlanta's south side. A couple of ladies initiated a conversation with Logan and me while my father spoke with his friends. Small talk and laughs were exchanged, and all was well. But 'the cat was let out of the bag' when it became clear to them that I was in the middle of an overseas tour and Logan was a recent combat vet. Things turned sour. The two gals became experts on foreign policy, lambasting Logan and me for serving in the military and being a part of the war machine. I'm not sure what it is with millennial ladies, but they persist in telling you everything on their minds, whether you want to hear it or not.

One of the young gals sipped her glass of wine and chirped. "I dunno why y'all are killing those people over there."

"Really, hon?" I asked, while seated at the bar top beside the ladies. "Are you seriously trying to have this conversation?"

"It's bullshit."

Although I sympathize with folks against the war, I couldn't sit by and allow the uninformed to badger Logan and me. "Look sweetie, you wouldn't understand, anyway. But no one is indiscriminately killing people."

I despised fact that I'd gotten sucked into this foolish conversation. *Night one, for real?*

"That's not what I see," the young woman said. "Every day you see reports where civilians are being slaughtered."

"That's the biased news," Logan chimed in.

"None of those people deserve to die," the girl said.

"Well hon, what do you suppose we should do when we're getting shot at? Just die?" I swigged my Miller Lite.

"They aren't attacking you," the gal answered. "You're the ones fucking with them."

"Oh, they ain't?" I didn't know whether to laugh at her ignorance or become enraged.

Ole Big Ronny B arrived in time and caught some of the conversation. With a cool head he said, "Let's go, Little Ronny and Logan. Y'all don't need to be having this talk with these gals who don't know shit."

"Whatchu mean, don't know shit?" one girl blasted back.

"No, y'all don't know shit, and we're just here trying to have a good time." Big Ronny B sipped his Miller and continued. "That's my son and I'm proud of him. I'm proud of both these boys. Whether you like what they've done or not, you should at least respect it. They didn't come here asking for your bullshit and judgement."

"I wonder how those poor Afghans felt when they were subjected to the US military's bullshit?"

"You know what, hon? I had to deal with people like you when I came home from Vietnam. Y'all weren't there, you wouldn't understand, and you're not worth anymore of our time." Big Ronny B had gathered the group and attempted to leave.

"Oh, so you're no better than they are," the gal said with hostility.

"You're right, hon. I used to shoot Vietnamese kids for target practice," Big Ronny B said sarcastically, trying to piss off the gals and move on.

We stood to leave, and bystanders approached. "Y'all come sit with us and have whatever you want. Thank you for your service and fuck those bitches!"

I vegged out for a day or two, and then borrowed my dad's car and drove to Wildfire's office. As soon as I parked and exited my car, Emmy flagged me down from across the lot. We met and hugged it out.

"Damn bro, good to see you," Emmy said.

We made small talk for a bit. Emmy offered updates on his situation. "Dude, you won't believe this shit, bro."

"Waz up, man?"

"I was just laid off, dude. The economy tanked, and price of scrap metal plummeted."

"No shit, man."

"Yeah dude, I had to short sale my house, and I'm living with a friend."

"Damn dude, it's rough on a cracka out here."

After a laugh, Emmy said. "Yeah, it's tough but you know me. I'll be back, baby." Emmy thought for a second and placed his chin inside his right hand. "You know, I wish I'd done what you did, Ronny B. You made the right call, bro."

It was then I understood I'd, indeed, made the correct decision. Yeah, there was bullshit. Yes, I was disillusioned with the war effort. I despised ROEs and undesirable positions we were placed in. But Uncle Sam was still the best place for me. Emmy went from a comfortable six-figure earner to being spit out by a corporation during a fragile economy. He was dedicated to the company and made it a ton of money for a long time. But the company wasn't dedicated to him.

Emmy grabbed me in a headlock. "Fuck my bullshit; let's go see Wildfire. Damn, it's great to see you. We're gonna have fun tonight. Ole Wildfire and I wanna hear about shooting Hajis!"

"Look who's here," Emmy announced as he led me into the office.

Fresh-looking artificial trees and plants sat in the corners of the waiting room and golden-framed hipster art was mounted on the walls. The cleanliness of Wildfire's business far exceeded that of his home.

"Oh my God, Ronny B," a young woman said from behind the receptionist's desk. "We were worried about you."

The entire staff hustled into the waiting room to greet me with hugs.

"Hey Wildfire, Dr. Jefferson, look who's here!" the young lady shouted.

Wildfire strutted into the room, grinning; Dr. Jefferson followed.

"Ronny B, baby!" Wildfire said. "Welcome home, buddy!" He grasped me in a bear hug and lifted me off the ground. "Man, you know I love you!"

Dr. Jefferson wrapped me in his arms, too. "My buddy is back. You know I missed you and prayed for you every night."

"I appreciate it, Dr. Jefferson."

"You and I need to have dinner."

"Will do, sir."

After the initial meet and greet, Wildfire led me to a patient room. "Lay in that chair, boy, and make yourself comfortable."

He scrubbed my face clear of blackheads and rubbed a gel on my cheeks. He also fired up a laser and methodically worked it over my skin. I had no clue what was going on.

I felt a needle being injected into my forehead. "Are you shooting me with Botox?"

"It's all good, Ronny B," Wildfire said. "You've got good skin. I wanted to get rid of a line on your forehead."

"Damn man, you're gonna have Daddy lookin' sexy."

"That's how I like you, sweet meat," Wildfire said, and we laughed.

Once procedure was completed, Wildfire led me to his office, where Emmy was seated behind a grand, antique-looking wooden desk. An entire wall of shelved books rested behind him. An air force basic training platoon picture was framed in a prominent place next to an autographed picture of George W. Bush.

"I ain't got no more patients today," Wildfire said, as he reached inside his desk drawer and pulled out a small white baggie.

Wildfire packed a bullet-looking thing and passed it around. "I wanna see some pics. I know you've got some shit."

I logged into my email account and downloaded some pictures that we projected onto a large flat screen on the wall.

"There's my boy." Wildfire stared at a picture of me on patrol. "Now let's see some real pics, Ronny B. C'mon man!"

Since Wildfire and Emmy were good friends, I showed the pic of the dead Haji lying in the room with his eyes wide open. The next picture popped up, and it was the one Fink snapped of me straddling the dude.

"Motherfucker!" Wildfire yelled; he was fired up from the blow. "Fuckin' A, that's my boy."

I noticed Wildfire saving the picture to his computer. "Dude, don't be emailing that shit to anyone. I don't wannabe on CNN."

"Who's your boy?" Wildfire asked as he clutched me around the neck. "It's all good with ole Wildfire, baby!"

I showed pictures of blown-up vehicles, the turret hit by an RPG, the back of my armored vest ripped by a bullet, and the rolled over MRAP that almost killed me.

"Holy shit, you're a lucky motherfucker, Ronny B," Emmy said, hitting a bump.

For old time's sake, Wildfire, Emmy, and I dressed in old skool seventies clothes to hit the town. With butterfly collars, polyester sportscoats, and aviator shades, we were stylin' and profilin'.

"Hit it, Ronny B!" Wildfire demanded, as he shoved the bullet thing in my face.

I took a few sniffs and passed it to Emmy. Wildfire swigged from a highball and stomped the pedal of his BMW convertible, blowing through the first red light. We were headed to Hotlanta.

Emmy and I followed Wildfire's lead. He led us around the snaking line of a popular nightclub and proclaimed. "I'm fucking Wildfire, baby." We breezed past the doorman.

After elbowing our way to some barstools, Wildfire ordered a round of beers and shots.

Wildfire still had his shades on which prompted a gal to say, "You look like you're in the FBI."

He stared down the woman with his patented smirk and said in a thick southern drawl. "Naw honey, but I was in the air force. That's my buddy there and he's been goin' gangbusters on ragheads overseas!"

As if a record had stopped, everyone shifted toward us. It was a yuppie joint. I nudged Wildfire. "C'mon man, what the fuck? Do you not have a filter?"

Wildfire laughed and said, "It's all good, Ronny B. Fuck 'em if they don't like it."

More shots were ordered and sent our way. The audience thanked me for my service and grew to love Wildfire's wit and charm. He happened upon a stranger who'd also served in the air force. The two must've talked for hours.

"I love talkin' that old shit," Wildfire said as he patted his new buddy on the shoulder.

I made my way back to my father's home the next day. After logging on Facebook, I noticed Brandon online. I initiated a chat to get the latest scoop from Afghanistan. Following some small talk, Brandon mentioned the platoon was leaving soon for Margah. They'd spend Christmas there. I wished Brandon luck and prayed for the platoon.

I made more rounds visiting friends. Wayne and I knocked back beers and reminisced about the good ole days. He opened a bottle of whiskey and chugged a massive gulp. I laughed and reminded him of that time in the parking lot at a Slayer concert. Showing off for the chicks, Wayne turned up a liter of whiskey and sipped three quarters of the bottle down. I'd never seen anything like it. I picked him up from South Fulton Hospital the next morning after he'd registered a .42 BAC. I wouldn't have believed it had the nurse not pointed out Wayne's blood alcohol content and insisted "y'all need to be more careful." Ole Wayne smiled over the memory, flashed the horns, and turned up Gary Stewart.

I couldn't be in the area and not pay my fav Uncle Hook a visit. He had few valuables but tons of wisdom; ole Hook was a fabulous storyteller. One of his popular stories was about being drafted. A sergeant had placed his foot on the scale to increase the wiry Hook's weight to the minimum to be fit for service. Hook wasn't the most compliant soldier, though. An officer busted him smoking pot inside the tool room. When he was called to attention, Hook often flipped a bird and told the officer to fuck off. He left the army and became a self-described "dirtbag hippie" and a "ramblin' man." Society may have considered Hook a loser, but he never gave a fuck what society thought. With a tallboy can of Busch in hand and a cig burning in his ashtray, Hook was a hoot to be around if you could stand indoor cigarette smoke.

Christmas arrived but I can't say I was feeling the spirit. Thoughts of the boys in Margah consumed my mind. I spent the day going through the motions as Big Ronny B, my sister Shelly, and I visited family. We dropped in to see fam from my mother's side and stuffed ourselves with a delicious Christmas dinner. I felt a bit strange being around cousins who were my age and had their own families. With their honest civilian careers, mortgages, security and all that jazz. They didn't make me feel strange; it was my own foolish thoughts of being behind. Nonetheless, the forces of family and Christmas joy overrode my worries of Margah.

Some old friends visited the south side of town during Christmas Break to see their parents. A friend from my neighborhood named Eric spent a night with the in-laws, who lived down the street from my father. Eric dialed me up for a drink. A proud Bulldog and graduate of the University of Georgia, Eric was a successful white-collar guy who did well. Since he was a staunch conservative, Eric was proud that I served. Whenever he said, "thank you for your service," he meant it. We agreed to meet at a nearby watering hole that was a mere walking distance. Since Big Ronny B liked Eric, and sometimes treated him as a second son, he came along to enjoy the night.

Rather than walking, Big Ronny B decided to drive the short distance due to his medical condition. His physical appearance hadn't noticeably changed, but the pulmonary fibrosis was making breathing difficult. The joint was packed, but we grabbed a seat to chitchat. It was good talking to Eric, and we reminisced about our childhoods in the old hood. We spoke about current events, how his life was going, and I confided in him what I thought about deployment. We lifted our glasses and proceeded to pound drinks since the night was low risk.

"How are dem Dawgs gonna be next year?" I asked.

"We'll see, they've gotta couple good quarterbacks coming in."

"Yeah? I keep hearing these Georgia fans going all gaga over this Aaron Murray kid."

"They've gotta another kid named Mettenberger arriving, too. He's big and physical."

I blacked out due to the number of drinks I guzzled. Details remain fuzzy. Eric and my father helped me to the car. Big Ronny B was buzzing but not shitfaced. The short drive home should've been an easy task. But things went sour when my father arrived in the neighborhood and attempted to drive Eric to his in-laws.

"Drop me off at your house, and I can walk to my in-laws," Eric said.

But Big Ronny B, hard of hearing, replied, "Yeah, I'm gonna drop you off at your in-laws."

"No, Big RB, I'm fine. Just drop me off at your place."

Big Ronny B was passing his own driveway when he understood Eric. "Oh shit, my bad. Lemme turnaround."

My father did a U-turn inside a cul-de-sac. He was driving toward his home when a cop pulled behind us and hit the lights. We coasted into the driveway. Big Ronny B had the gift of gab and handled situations well, never providing direct answers to questions.

The officer approached, and Big Ronny B spoke. "My son is home on

R&R from Afghanistan, so we decided to celebrate. I drove since I have trouble walking."

The policeman asked if he had any alcohol to drink as other officers arrived on scene. Red lights blazed the neighborhood.

Big Ronny B laughed and answered with a question. "Well, what'd you think we were gonna do while he's home from the war? Sit home and stare at the walls?"

The cops couldn't help but laugh, and one asked, "Do you think it was a good idea to drive tonight, sir, knowing y'all would be celebrating?"

Big Ronny B, cool as ice, said, "But, I'm already home."

Everyone chuckled.

My sister was over for the night and vouched for our father's story. The cops understood the deal. Shelly scolded Big Ronny B and me while Eric helped me inside the home.

Time was wearing thin, and I thought it'd be nice to be in the company of a sweet young lady. *Should I message some ole gal who'd been flirting with me online?* As I thought of a last resort, I ran into a couple of gals inside a coffee shop. We recognized one another from a local sports bar. These were young, good-timing ladies who were down for anything. In no time at all, I was seated inside their apartment, smoking a blunt and drinking wine. *Hot damn.* After a full day of nonstop partying, I called Hix to fuck with him. He'd gotten injured again and was sent back to Alaska. His war was over.

"Hello."

"What's goin' on, baby?" I asked, in a haze.

"Ronny B, what's goin' on, man?"

I put the gal on the phone, and she cut in. "Hey honey, this is Kathy."

"This is who?" Hix asked.

I spoke up. "This is my friend Kathy; tell him who you're with, darlin'."

"Oh yeah, I'm with the colonel, baby."

As Hix laughed, I asked. "How old are you, sweetie?"

"I'm twenty-four," Kathy said, as we had rehearsed. "I'm young and sweet, too."

"So, you're sweet?" Hix asked.

"As candy, sugar."

"What are you guys doing?"

"The colonel's showing me what a real man is all about."

Laughter spewed from all sides. I grabbed the phone. Hix filled me in on his condition and what was going on in Alaska. After a few minutes, I let him go because sweet Kathy grew feisty.

Kathy loved to party. But despite being young, she had her shit together. Armed with an education, nice job, independence, a little maturity and drop-dead gorgeous looks, Kathy was a boss woman. Kathy wasn't going to call me Daddy, and that was okay, because I was chilling in the company of a beautiful young lady.

I couldn't help but hear the melodies of ole Bocephus singing *Outlaw Women* when Kathy snorted a massive line of cocaine. Following a bump, she'd swig from a bottle of Jim Beam. In no time at all, we were wrestling around naked, filled with energy and extra horny. I applied a hammerlock, pulling her arm behind her back and tweaking it upwards towards her shoulder blade. Kathy twitched and let out an ear-piercing scream. I released the hold; she shifted and smacked my shoulder.

The wrestling progressed into something steamier. Kathy laid a passionate kiss on me while straddled atop my body. She was in a dominant position—a position I gladly accepted. She nibbled on my neck and my manscaped chest.

I flinched when she bit my nipple. "Naw darlin'." Kathy chuckled and realized how ticklish I was.

She continued massaging my chiseled body, a body she adored. The endless mountain hikes, beast mode gym workouts, and recent steroid cycle were paying off in the bedroom. I grasped ahold of sweet Kathy and placed her into a position she loved. She relinquished control as I gently kissed her tight, fit figure. A body bump of blow was laid out, and I snorted from the top of Kathy's chest to the peak of her perky nipple. We exchanged body bumps. What a helluva way to celebrate ole Ronny B's march home.

R&R was going according to plan. I met the people I wanted to see. For the most part, I'd let the war go and morphed into relaxation mode. But I couldn't get the thoughts of my boys in Margah out of my head. One evening, I watched the world news and a horrifying story concerning Afghanistan rocked me. A suicide bomber, Khalil Abu-Mulal al-Balawi, made his way into a CIA facility on Camp Chapman and detonated an explosives-laden vest. Ten people were killed, including al-Balawi. Seven of the ten killed were CIA officers and an additional six CIA officers were wounded. That made this the most lethal attack against the CIA in more than twenty-five years. It was December 30, 2009, and I was flying back to Afghanistan in a couple of days.

I was scheduled to be at the airport on New Year's Day. But I figured I'd stretch it another evening. Doc Chauncey was on the same R&R rotation and made it clear that he wouldn't be returning to the airport until January 2. Since Doc Chauncey was a well-respected leader in Charlie Company, I felt assured. Flights departed Atlanta and Dallas for Kuwait daily, sometimes more than one per day. I'd have no problem catching a flight and meeting the fellas in Kuwait or Bagram. I met with the crew at Taco Mac to ring in a new decade. Big Ronny B joined the festivities. Mark and Murph, two old skool ClayCo friends, arrived along with a slew of people who drank too much and partied too hard.

I drank too much, myself, since it was my last hoorah before boarding the anti-freedom bird. Friends and bar patrons doled shots as the pride and patriotism flowed. They wanted to send their own Southside ATL boy off right. I guess the party was 'all that and a bag of chips' because I awoke on a couch inside a hotel lobby. It was four in the morning, and no one was around besides the graveyard employee working the front desk. My head was smoking. While attempting to shake it off, I tried calling a cab. There was no answer. I walked across the street to a Waffle House to grab a bite and figure out how to improvise, adapt, and overcome the situation.

I sipped coffee, desperate for a ride to my father's home. I noticed an African American dude about my age eating alone so I offered him money for a ride. He agreed, so we finished our meals and headed for the door.

As we approached his vehicle, I recognized the eagle, globe, and anchor sticker on his back windshield. "You were in the marines?"

"Yep," the man said. "I was an artilleryman at Twenty-Nine Palms."

"No shit? I was a combat engineer on Okinawa and Camp Pendleton."

"My dawg," the man said and smiled as we sat inside car. "Good to meet you, devil dawg!"

"Semper Fi, jarhead."

The brother reached inside his ashtray and pulled out a half-burnt blunt. "Wanna smoke some of this?"

"Do I?"

We passed the chronic back and forth as we cruised down the road. I confided in the jarhead about my current situation as the cannabis hit my brain. We laughed and talked for the duration of ride.

Once we hit my dad's driveway, we smacked hands and the gentleman said, "Be careful over there and take care, dawg."

"Thanks a lot, man."

I tried to hand him a little cash, but the brother said, "Naw dawg, you don't owe me a thang, baby."

"Semper Fi," was the last words we exchanged before the ole jarhead drove off.

On January 2, 2010, I was dressed in uniform and standing erect inside the ATL Airport. Big Ronny B slapped me with a hug. "I love you, son. Don't try to be a hero. And don't be afraid to pull the trigger again."

I fought hard to keep the tears back. "You know I won't. I love you too, dad."

I turned to rejoin the boys in green. It was high time to get back into warrior mode. *Six more weeks; let's do this, baby!* As I continued my walk through the terminal, a peaceful feeling came over me. I was satisfied with my short trip home. Nine straight months of saving and not spending money allowed me to use a portion of my nest egg to "ball out" a bit. Big Ronny B and I had peeped a Hawks' game in style, inside the VIP wing at the Phillips Arena. We sipped fine bourbon and watched the Hawks take on LeBron James and Shaquille O'Neal with the Cleveland Cavaliers. I lived like one of my idols, the American Dream Dusty Rhodes. I lived those two weeks "on the end of a lightning bolt" and was certain Big Ronny B did the same during his two weeks of R&R in 1967.

When I reached the boys in green, the rush returned. I was transitioning from temporary ATLien baller to a Dedicated Infantry Combat Killer. I had a bit of a pep in my step and thought about my final encounter with Kathy.

She had massaged my shoulders as we embraced. "I fucking love you, Daddy." She had laid her juicy lips onto mine and said, "I hope my War Daddy stays safe."

But I was also eager to return to "the 'Stan" and see my boys. I was itching to get back on the manifest. No matter how shitty the game, no participant enjoys sitting on the sidelines. Time to finish the drill, baby.

CHAPTER 15: THE FINISH LINE IS MARGAH

The bird hovered on the Boris HLZ. I stepped off with my gear and hustled into the FOB, ready to complete my tour. Only six weeks remained. I'd convinced myself the platoon would wind it down with little drama. I strolled to the cans to reclaim my bunk and get reacquainted with everyone.

"Look who's here," Joey C said. "Ole Ronny B, baby!"

We clutched one another in a tight bro hug. "Good to see you, man."

I made the rounds before heading to Joey C's room to smoke a cig and shoot the shit. Brandon and T-Love sat on a bunk.

"Man, I'm ready to ride these last few weeks out and be done with this shit," I said.

Joey C stood silent. Brandon pointed to a paper calendar with events penciled in.

"Well, hate to break it to you, Ronny B, but in a few days we're gonna run a mission every other day for three weeks," Brandon said. "Then we spend our last two weeks at the Margah compound."

I felt as if I'd been kicked in the balls by Bruce Lee. "Fuck my life! What are we trying to do? Win the fucking war in the last few weeks?"

"Oh, by the way, since you've been gone, we gotta new lieutenant and platoon sergeant," Joey C chimed in.

"Oh yeah?"

"Yep, Lieutenant Hart was the XO of one of the other companies and Sergeant PK was moved from headquarters."

"Ah, I see." I gritted my teeth. "That lieutenant never went "outside the wire" as XO so now he's trying to get his."

They told me about our new leaders. Lieutenant Hart was a fresh-faced West Point grad. He appeared soft-spoken and deferred to NCOs for guidance. With larger than average ears, dark hair and eyes and a boyish appearance, Hart wasn't intimidating. On the other hand, Sergeant PK was an experienced E-7 who'd deployed several times. He'd been inside numerous battle spaces throughout Operation Iraqi Freedom. Despite PK's experience with combat, he was approachable and had no problems shooting the breeze with lower enlisted guys. His leadership style was a complete one-eighty from Manny Fresh.

Also, while I was gone, two of our MRAPS were fitted with the CROWS (Common Remotely Operated Weapons Station). This system allowed a gunner to sit inside the back of a vehicle instead of being exposed in the turret. Gunners operated the system by viewing an area on a video screen and controlling the weapon with a joystick. Every crew-served weapon system, .50 cal, Mark 19, and 240B, operated with the CROWS. The human element of aiming and firing was removed and replaced by a computer's pinpoint accuracy. It was the perfect tool for millennial soldiers who grew up playing video games. The CROWS was tested when a firefight erupted from the Margah compound during the Christmas rotation. Hotdog-sized .50 cal rounds had ripped through the attackers. Paris had picked off several insurgents while gunning the CROWS, further feeding his addiction to the infantry.

We rolled out every other day as planned. I served as a gunner in Rubio's vehicle for most of the missions during the homestretch.

He'd peek upward and say the usual. "Hey Bruce, if something goes boom, find something, and shoot it."

"Right on, sergeant."

Sometimes I served as a dismount and paced through villages, as I had a hundred times before. Often, indirect rounds would impact around our position, prompting us to take cover by kalats. During the wintertime, snow often covered the ground, but it didn't prevent us arctic warriors from patrolling. It's said many casualties occur during the tail end of a deployment. So, we exercised keen situational awareness, not wanting to become a victim.

In the past, Big Ronny B mentioned that he never slept during his last days in Vietnam. "The Vietcong snuck up on guys at night and killed them while they were on guard duty."

After completing the scheduled missions, we packed our bags and rolled to Margah. We emptied our rooms and gathered all our belongings since a bird would scoop us up from the ABP compound and fly us to safety. Margah was the finish line. Since Stan was "stop-lossed," the lifers cut him some slack. The unit plucked all the stop-loss soldiers two weeks early and sent them home. Stan gathered Second Squad to say goodbye and offer a heartfelt plea.

"Don't get complacent," Stan said, while stuffing his rucksack with gear. "Stay alert and be aware of everything around you. Many grunts get killed in the last days before they're supposed to go home. So please watch your asses and look after one another."

The squad said its goodbyes and, just like that, ole Iron Guts was gone. The platoon was also without two of its original squad leaders, Stan and Hix. But we'd grown during the past year and were groomed for the task.

Second Platoon's convoy was ready to roll out when we arrived in Margah. They were worn down and eager to "peace out." Being inside that miserable place made me nervous too. A nine-man element pushed to the top of the hill where the COP once stood. The rest of us remained on the bottom to man the compound. It was mid-February, and we hunkered in to finish the drill.

Our time at Margah was very different than our compatriots battle at Marjah in Helmand Province. They grabbed headlines when they launched their assault on Marjah. These history-making "Marjah Marines" had cameras, documentaries, and notoriety. Without a doubt, they possessed the full power and fury of Uncle Sam's arsenal. They'd initiate contact and seek all the advantages we grunts are supposed to hold. Yet here we were sucking dick for beer money in Margah. No lights, no cameras, but plenty of action.

A BBC reporter with a pair of stones provided the world with a glimpse of Margah in 2011. He proclaimed while reporting from there. "This is the most attacked base in Afghanistan."

During the boredom inside the unfinished building, I read several magazines. It freed my mind of anxieties. I stumbled across a book of Hunter S. Thompson stories as thick as a phonebook. Since I'd seen *Fear and Loathing in Las Vegas*, I dove into the pages. One story that stole my attention was Thompson's attendance at a speech given by Georgia Governor Jimmy Carter in 1974. It'd become known as the *Law Day Speech.* Thompson had sipped his glass of Wild Turkey and sat amongst Georgia's elite lawyers and judges. This little-known redneck governor had captivated Thompson as Carter scolded the attorneys over imbalances inside our nation's justice system. Thompson's newfound admiration for Carter helped lead to a Carter presidency. It was a good distraction from reality.

We continued patrolling the bazaar and villages during these last two weeks. Shuras were conducted, kalats were searched, and storefronts inside the bazaar were explored. Shelling commenced as soon as we stepped off the compound or whenever we reached our destination. Enemy spotters eyed our every move. I carried a radio as a team leader and listened while an interpreter interrupted several times to tell us what spotters were saying. I wished I was oblivious to everything happening around me, as I was before a radio was issued to me. It'd be one less thing to worry about.

Tensions grew during one patrol inside the Margah Bazaar. During early afternoon, the platoon embarked on a short walk from the compound to the bazaar. Stu and I placed our guys at the corners to provide security as leadership conducted its business. The bazaar was a short strip of dirt road with spooky-looking wooden buildings along both sides. It reminded me of an Old West gunslinger town like Dodge City from *Gunsmoke* – but a Stone Age version of Dodge City. Stu and I wandered behind the decrepit buildings as we checked on our boys. The junk, trash, smell, and funk would make a billy goat puke. As we scanned through the rubble, I noticed a dead puppy lying on its back with its eyes closed and mouth wide open.

Stu stared with a stone-cold face. "Man, this is a different kind of place."

Crack, crack, crack sounded when automatic gunfire interrupted Stu. We hopped into the rubble behind the building to take cover and assess the situation.

"Goddamnit, it's just D-Rum firing warning shots," blurted someone from the area.

Ole D-Rum rattled off a few rounds into the ground below a motorcyclist who was approaching a little too close. Lieutenant Hart wasn't pleased, but I didn't blame D-Rum. Fuck these people and their feelings; better to be safe than sorry.

At the compound. we continued with the same silly conversations, card games, and guard duties. One evening, I stepped out to the CROWS MRAP to relieve D-Rum from guard.

I was met with a smokey surprise when I opened the back hatch. "Holy shit, it smells like Cheech and Chong are hangin' in here."

D-Rum, kicked back and red-eyed, stared at the monitor. He rapped some original lyrics and appeared to be "in the zone." But now wasn't the time nor place.

D-Rum took a drag, exhaled, and said, "Yeah man, this is some fire-ass shit that ole ABP dude has."

"Damn, man, why don't you open the top hatches and let this motherfucker air out?"

"Fuck, I dunno, dude."

D-Rum handed over the joint and left to get some sleep. "Dude, have this killer shit; you won't be disappointed."

Entering the vehicle, I opened hatches to rid the stench of marijuana. I dropped the left-handed cig inside a water bottle and hurled it as far as I could. The CROWS was too complicated to operate while under the influence. I'd never be able to live with myself if something went down during my shift. Thoughts of Stan hammering complacency and Big Ronny B staying alert during his final days in 'Nam consumed me.

There were a couple of days of silence. No mortars, no small arms, no shots fired. I was content with things staying that way. Yet one day as the afternoon sun faded, a warning came through.

An interpreter rushed into the makeshift TOC, trembling from nerves. "They're about to fire seven rockets at us. They're prepping the rockets as we speak!"

Hart and PK explored their options. They radioed guys on the hill to see if they had eyes on anything suspicious. A group of men were spotted prepping rockets to fire from the back of a pickup. The enemy was out of range for crew-served weapons, so Lieutenant Hart dialed the company TOC to request a fire mission. But like so many times before. "Denied."

"Motherfucker!" I shouted with disappointment, as we sat helpless inside the compound.

"Unreal." Gus sat on a cot across from me with his head in his hands.

"Do they want us to fucking die?"

Gus lifted his head. "If a round lands on this building, we're all smoked."

"And we could put a stop to it right now."

But there was no time to wallow. Despite his disappointment, Sergeant PK spun into action and rallied the platoon.

"Bruce, grab this .240 gunner and head to that tower. Gus, get your SAW and take these two with you and go to the tower to the east." PK paced the room, gesturing with his hands. "Mac, grab a couple of guys and set up at the entrance. Archie, see if you can dial air support."

I wasn't crazy about sprinting under incoming fire with a .240 gunner. There was a method to the madness, though. PK was doing the best he could under shitty circumstances. An NCO from mortars approached PK with a plan. Dusty, a native Coloradan, was never afraid to voice his opinion.

A cigarette burned between Dusty's lips and a Haji rag draped his neck. "Why don't you call the TOC and see if they'll let us fire illumination rounds over their heads? It's our only chance."

While darkness settled and the clock ticked, Lieutenant Hart dialed the company TOC and asked to fire illumination mortars. Five mortars were approved and launched over the insurgents' heads. Brightness lit the night sky.

"Oh shit, they see us," was interpreted through the radio, and the pickup sped away.

The rocket attack was thwarted, but I fumed inside. An enemy went unpunished after trying to kill us. *We let 'em live to fight another day.*

With only three days left in-country, an element of the platoon patrolled west towards Volkswagen. A call over the radio informed us of enemy spotters tracking our movement. We allowed a platoon of ANA to lead the way to relinquish some control. An enemy AK 47 ripped off several rounds. I hit the deck. No structure existed to take cover since we were walking across the bare flats. Bullets skipped into the dirt near Joey C and Brandon, and rounds fired from kalats to our front. One thing to be thankful for was the ANA being positioned in the lead. At least we Americans possessed the discipline not to shoot with friendlies to our front.

Once the dust settled, Lieutenant Hart radioed Rubio. "Three-Three this is Three-Six, we're gonna push into village and take a look."

Fuck my life! Hart was itching for a firefight with three days left. I sprinted to Joey C and Brandon's position to pass the news.

"We're gonna push into the village," I said.

"Really?" Joey C asked, lying prone in the dirt.

"Yep, let's do this and get the fuck on."

After informing my team of our orders, an ANA Ford Ranger sped by toward the compound. Their first sergeant had shot himself in the foot by accident while attempting to return fire. *What the fuck?*

We picked up and pushed into the neighborhood. I linked with The Duke, and our teams climbed onto a roof to provide overwatch for the element searching on the ground. The hairs on the back of my neck tingled. Leaving no stone unturned, I searched for insurgents through my optic. This kalat had a pimping deck that doubled as a roof; it provided sufficient cover and excellent sightlines. Seconds lasted hours until the call came over the radio to pack up and hike home.

The next day, Second Platoon drove a convoy out to chauffeur our replacements to the compound. This new bunch called themselves Rakkasans. They strutted through the area wearing clean uniforms and high-speed gear. I sensed the Rakkasan boys eyeballing us as we sported torn and dirty uniforms, emaciated bodies, and scraggly unkept beards. It brought me back to a year ago when we exited the bird at the Malekshay HLZ and stared down the battered group of men we replaced. Now we were those dudes: the walking dead hobbling around with thousand-yard stares.

Lieutenant Hart instructed the new lieutenant to send nine guys up the hill to provide overwatch. I peeped as the men stepped it out up the steep hill. They bent over and gasped for air several times; my first walk up Malekshay came to mind. It had kicked my ass and put the elevation into perspective. We were told to dry shave that evening since the incoming and outgoing brigade commanders were flying in. I was pulling radio guard inside the makeshift TOC when both colonels arrived to peruse the area. The incoming colonel crossed his arms and sneered while browsing the big shit sandwich being handed to him.

Our colonel explained rotations for Margah as the Rakkasan colonel's lip curled. "From my experience, a place like this isn't sustainable for a year." Rakkasan-Six was pissed.

The following morning, the order was unexpectedly given to get our shit together. A helicopter was landing in thirty minutes.

"Holy shit, Joey C, we're gettin' outta here a day early!" I shouted and pumped my fist.

Ric Flair-style "woooo's" blanketed the room.

We lugged our gear outside the south entrance of the compound and waited inside a clearing. Pure joy mixed with gripping anxiety. I stood in the open, thinking about a possible rocket attack or one final ambush, until the roar of helicopter rotors filled the air. My heart throbbed. Third Platoon rushed onboard the helicopter, screaming as if we were inside a mosh pit at a metal concert. We lifted off and a feeling of elation rattled my body. I gazed across at Joey C who laughed out loud, nonstop, with a smile glued to his cheeks. I leaned back like a gangsta and flashed the horns as the bird breezed across clear blue skies. I savored the moment, as young men surrounded me, basking in what might have been the greatest joy of their lives. *The war is over for me.*

CHAPTER 16: THE ALABAMA FREEZER

"Wow, you guys are looking good," the sweet young gal said to Wimes and me while she gripped a beer.

"Stylin' and profilin', darlin'." I let loose a Ric Flair line.

"You guys are dressed nicer than anyone here." Wimes and I were rockin' fresh tailored sportscoats.

"That's cuz we're not Alaskan trash, dear," I said and winked.

"Oh no, you didn't."

"Oh yes, I did." I toasted the beer in her hands.

"I can tell you're not from around here." She took a step towards me and licked her lips.

"Naw, honey."

"How do you like it?"

"I dig it." I took a step closer, bringing us nose to nose. "If you take Alabama and put it in the freezer, this is what you get."

Wimes interjected, "Holy shit, we're in an Alabama freezer!"

Damn, it's great to be back inside the Alabama freezer!

Everyone was excited to be back in Anchorage. But this time, things were different. Some guys had fulfilled their obligation and left the army after the tour. Others moved to different units, and some of the original squad leaders who remained were shuffled to another company. A group of new guys checked into the unit while we were deployed. We didn't treat new guys like shit; we were indifferent. Most of us who remained, though happy to be home, were drained from the deployment. The new guys didn't get our best welcome. When most of the NCOs were removed from the platoon, it broke our backs. Since we didn't sweat and bleed with the new NCOs and privates, it was less of a brotherhood.

Problems gripped the men during the next few years. Suicide, alcoholism, drug abuse, and medical issues plagued the entire brigade. The road to an honorable discharge became long and winding. It didn't matter if a guy was decorated, an experienced NCO, or a Purple Heart recipient, soldiers who made poor decisions were kicked to the curb. They were flown home with a document reading "something lesser than an honorable discharge." Things got more complicated when these guys returned to their hometowns. It's hard to rebound from a "big chicken dinner" (bad conduct discharge) served by Uncle Sam. VA benefits are suspended, and job prospects are grim.

Before the deployment, some guys had records: DUIs, possession of narcotics, positive drug tests, domestic violence complaints, and an entire slew of legal issues. Originally, command wrote off these misdemeanors, and the soldiers were shipped off to war. After the deployment, however, if a soldier stumbled, they were "shit canned." Bodies were no longer needed as the Iraq War drew down. Since the army plays a numbers game, the easiest way to cut them back is to start with people who find trouble.

Within a year after the deployment, a couple guys from the battalion took their own lives. Staff Sergeant Stansberry was a solid dude from Arizona in charge of the small motor pool on FOB Boris. I'd stop by and shoot the breeze with him and the other mechanics. He struck me as a laidback cat who'd chat with anyone. Staff Sergeant Stansberry shot himself inside an Anchorage carwash. He was thirty-three.

The other soldier from the battalion who died was a medic and Doc Lindley's roommate in the barracks. According to Doc Lindley, the two were watching movies when Lindley decided to stay up late. His roomie passed out on the bunk. Doc Lindley snacked on potato chips and finished another movie. Then he decided to check on his sleeping roommate. There was no pulse. It was later determined the soldier died of an overdose.

Despite these heartaches, the men rolled along. The brigade of wound-up soldiers was released upon the city of Anchorage. During the weekends, the bars were slap full of partying soldiers. The barracks was full of young guns playing beer pong on foldout tables in the hallways. I was neither above nor beyond taking part in the wild drunken nights, but I managed to maintain my balance. Uncle Sam wasn't feeding me a "big chicken dinner" if I could help it. There were a couple fistfights in Anchorage that I wasn't proud of. One occurred inside the redneck Last Frontier's parking lot after I'd been boozing with Blackie. While exiting the bar, I witnessed a dude pissing near my vehicle's driver's side door. I grabbed the stranger by his shoulder, spun him around, and decked him in the face. His pants were unbuttoned, and his dick was still in his hand.

"The fuck you doin', man?" I screamed as I tackled the guy and pounded away.

Blackie cheered. "Get 'em, Ronny B; whoop his ass!"

The bouncers broke it up, but not before I made him pay for being a "tool." As Blackie and I were about to leave, the dude charged and smashed me square on the nose.

He ran off yelling. "I called the police!"

"Dude, you sucker-punched me, and you're calling the police?"

Blackie and I darted off in separate directions. I sprinted into the woods behind the bar and took cover behind a tree. After an hour, I resurfaced at a gas station and plotted my next move.

"Hey Ronny, that you?" a gal's voice said from behind.

I turned and recognized the woman. "Tamika, waz up, girl?"

Tamika was a large black woman who worked the door of the Last Frontier. She had the hots for me and always let me in free.

"I'm kinda in a jam, hon."

"I heard all about it, Ronny B." Tamika grabbed my arm and led me to her car. "Here, come with me."

She drove me to her place. After more drinks, I awoke in bed with Tamika. My clothes were still on.

"I could've taken advantage of yo' sexy ass," Tamika said while rubbing my chest.

"I know, dear." I laid a playful smooch on Tamika's lips. "Thanks for taking care of me, hon."

Sergeant PK called while I was with Tamika. *Oh shit!* "Yes, sergeant."

"Hey Bruce, you need to contact the Anchorage Police Department because they called looking for you. The fuck's goin' on?"

"I dunno what's up, sergeant," I said, playing dumb. "I'll call 'em."

I dialed the Anchorage PD, nerves pulsing, not knowing what to expect. The dispatcher said, "You're good, Mr. Bruce. We don't need anything."

The following Monday, I moped to formation. *Maybe everything slipped under the rug.*

Captain Mac exchanged salutes with me, and he said. "I heard about what happened over the weekend, Bruce."

"Oh yeah, sir?"

"Did you know that guy was in the air force?"

"No, sir."

Captain Mac leaned forward and rubbed his chin. "Did you kick his ass?"

"I'm not sure, sir." I attempted to downplay the incident.

"You either did, or you didn't."

"Yessir, I whooped his ass, but . . ."

"Good job." Captain Mac interrupted and patted my shoulder. "Fuck an airman."

"Fuck 'em, sir," I said and flashed the horns.

The fact was, I was disappointed in myself. I hadn't been in a fistfight in over a decade. Years ago, I was involved with some vicious brawls, but I fled that life.

After I'd been arrested for fighting (the charges were dropped), Big Ronny B once asked, "That road leads to death or prison; which one do you want?"

I felt better when walking away from fights. I've avoided them to the point where hands have been laid on me. I thought back to before deployment when I was hanging at Koots and chatting with a gal while we waited for friends.

A young white dude approached and shoved me in the chest. "You're talking to my girl, asshole."

Without addressing the punk, I turned to his girlfriend and said, "You've gotta jealous boyfriend, honey," and walked away.

The guy expressed how he'd kick my ass. "I believe you, buddy." The more nonchalant I appeared, the angrier he got.

He huffed and puffed. "You don't want none of this, bitch."

"Son, you're not worth rufflin' my hundred-dollar cashmere sweater." I stood my ground and called his bluff. "Now get the hell on, boy."

The punk acted as if he needed to be restrained, so the bouncers intervened. I waved bye and blew him a kiss. The ladies took notice of how coolly I handled the situation. Fast-forward to after deployment, and here I was fighting like a twenty-two-year-old with a temper. *What am I doin'?*

Lots of Charlie Company guys ran into trouble. Doc Chauncey pulled a knife on a cabbie, which resulted in criminal charges. He got booted. *Our hero medic did what?* Domestic violence, assaults, and batteries were reported. I drove to someone's home to bring them to the barracks because I was told over the phone – "I'm gonna kill this bitch if you don't come get me." There were alcohol-related hit-and-run accidents, guns drawn, and guys on the lam because they'd escaped police custody.

I've been awakened in the middle of the night and asked – "can you help me get him under control?" I walked into a barracks room to see more blood than the inside of an MRAP filled with wounded guys. It splashed walls, décor, and the floor. The dude had beaten the concrete walls with his fists until snapping out of whatever it was. A heroin junkie knocked on my door in the early morning and begged me for money to obtain a hit. DUIs and drugs ran rampant, and the army's method for dealing with these problems was to kick the soldiers to the curb. Maybe society could tame the monster Uncle Sam helped create.

After Wimes and I'd been out one evening, we caught a ride back to the base rather than risk driving drunk. All seemed well when we hit the third floor of the barracks. Minutes later, I heard a loud commotion. I peeked outside D-Rum's room and couldn't believe my eyes. Someone had a handgun drawn at point-blank on Brandon's head. The two stood in Brandon's doorway.

Brandon, a take-no-shit type with a touch of liquid courage, asked. "Whatchu gonna do, bitch?"

He was my squad mate, and I wasn't going to leave him hanging. I scored a large ax D-Rum had in his room and crept up the hallway. I took a position behind the dude with the brandished weapon, gripping the ax low. I was trying to be low key, but I could strike quickly if needed. The gunman focused on Brandon.

"I just called the police," D-Rum screamed from his room.

The dude with the gun fled. MPs arrived on the scene, put the barracks on lockdown, and investigated.

"Why are you holding an ax?" an MP asked.

"The dude had a gun." I nodded. *What the fuck?*

"Drop the ax now!"

I bent and placed the ax on the floor.

"Now, answer my question," the MP demanded. "Why were you holding an ax?"

Irritated, I opened my hands and shrugged my shoulders. "The dude had a gun. What don't you understand?"

"Turn around and place your hands behind your head."

"Are you kidding me?" At this point, I didn't give a fuck about their power to detain me. "Y'all received a call about a gunman, and I was trying to protect myself and my friend."

The MP moved to handcuff me, and I said. "The fuck was I supposed to do? This ax ain't shit next to his gun."

These dumbasses were giving me the biz and convincing themselves there was no gunman, that the call was a drunken hoax. We begged MPs to search the entire building because they were about to halt the search and declare "mission accomplished."

"You expect me to go about my business while a dude with a gun, who just pulled it on me, is on the loose in the bees?" Branded asked, expressing his own *what the fuck* look.

Another witness knew the gunman's name and room number. MPs found the dude inside his room and discovered his weapon during a second search that almost didn't happen. Guns aren't allowed in the barracks, so the guy was screwed, even if it couldn't be proven that he pulled it on Brandon.

"Y'all gonna take these cuffs off?" I asked with relief.

"We're gonna have to take you guys in to make a statement," the MP said. "But we have to place the rest of you in cuffs, standard procedure."

"Standard procedure?" I asked, squinting again. "Since when is it standard procedure to place cuffs on the victim, the good Samaritan who called 911, and the witness?"

"Standard procedure."

"This is bullshit."

Brandon and D-Rum were also placed in cuffs, and we were shoved into a squad car. We gave statements. Our story was corroborated through the investigations, and the gunman was charged and convicted in a court-martial. Brandon and I were issued a subpoena to testify.

Following the incident, the first sergeant pulled me aside to chat. "You know why I want to talk to you, Bruce, don't you?"

"I believe so, first sergeant." I stood at parade rest.

"Let's not beat around the bush." The first sergeant stared deep into my eyes. "You were with Sergeant Rubio when he got arrested downtown. You were also there when Brandon had a gun pulled on him, and you were with Blackie when he flipped out on a cabbie."

"Yes, first sergeant."

The first sergeant smirked oddly. "You're treading on thin ice, and people know your name. You've done nothing wrong, but you seem to always be around the fire."

"I understand, first sergeant."

He said something hardened infantry leaders rarely do. "If you need help, go get it, son. Go talk to somebody. I don't wanna see you in trouble."

The first sergeant was a good man. It wasn't his decision to kick soldiers out after they found trouble. But he understood what was being passed down from higher commanders. I heeded his warning.

I spent more time with sweet young thangs rather than blowing it out with the boys. Sweet Jane had moved on during my tour; the world still turns. Since we never committed to each other, it was cool to move on.

One gal was only twenty-two when I met her; she said. "I dig the older guys."

While lying in bed one night, flipping channels, I said. "Hey, that's Burt Reynolds," as *Smokey and the Bandit* was on.

"Who's that?" the gal asked with her head resting on my chest.

She had no clue who the superstars of the seventies and eighties were, which I found amusing.

Not all my encounters with women were humorous, however.

One morning I awoke, and a gal asked. "Do you remember what you did in the middle of the night?"

"No, waz up, hon?"

"I woke in the middle of the night, and your hand was on my throat."

"No way."

"Yeah, but you weren't squeezing or anything. You rose up, your eyes got big, and you said, 'Get down.' I said, 'I'm okay, honey.' Then, you raised your voice a bit and said, 'Get the fuck down!' I moved your hand off my neck, and you rolled over and snored as if nothing happened."

"I'm sorry, hon, but I don't remember that."

"I figured you wouldn't." The gal kissed me on the lips to reassure me. "It wasn't that bad."

Am I losing my fucking mind?

The old Afghanistan boys dropped like flies during the couple of years in the Alabama freezer. Many didn't reach their ETS (Expiration Term of Service), and the army washed its hands of them. It was infuriating. Good men were treated like bums and sent home. I'd wager a small fortune that at least a third of the Charlie Company guys who pounded Afghan dirt were bounced out of the service. Should people get kicked out of the military?

Sure, but for a heinous crime. Shit-canning people for one DUI, one failed drug test, or for any criminal misdemeanor is shameful. Especially when those were non-issues before deployment. A lot of people, including me, had trouble reintegrating back into society. Uncle Sam was taking no responsibility for the messes made. He passed the buck.

CHAPTER 17: FIGHT THE POWERS-THAT-BE

It didn't take long to realize the physical wear and tear from combat was catching up to me. Before my deployment, six to ten-mile runs were a breeze. Afterward, I was in constant pain. It took all I could muster to complete physical activities. We'd gather inside the Geronimo Gym to wrestle and train in basic ju-jitsu, which I enjoyed. Yet I'd walk away in pure agony. Pain is part of a grunt's game. You dance a thin line between what's acceptable pain and when a doctor should be asked to intervene. When I couldn't move one evening without icepick-stabbing pain, I sat in sick call the next day. Doc Lindley requested x-rays and MRIs after a thorough examination. Uncle Sam doesn't approve MRIs too often.

The results read that my knees and spine were toast. Doc discouraged me from partaking in strenuous physical activity. Doc referred me to physical therapy and set up trips to the Alaska Spine Center each month. This kicked off an exhaustive Medical Evaluation Board (MEB) process to medically separate me from the military. I could no longer train, so I could no longer be a grunt.

It sucks when you're no longer able to participate. You're considered a "broke dick" due to your physical limitations. There's a place on each base designed for soldiers dealing with injuries known as the Warrior Transition Unit (WTU), but the one on Fort Richardson was full. The WTU's mission is to allow soldiers to heal so they can either return to active status or separate. It also aids with the transition to civilian life. Instead, I'd remain with the company whose primary focus is to train soldiers and ensure they're combat-ready. Transitioning "broke dicks" to the civilian world wasn't a priority. Stu and Gus found themselves inside the same boat since both were also deemed unfit for service. Stu limped around on damaged knees, and Gus dealt with spinal issues. It was a relief having those dudes to lean on.

I stuck to a strict plan of physical therapy, prescription meds, and various treatments. I learned that exercise was necessary for alleviating back pain because doing nothing leads to larger problems. But strapping one-hundred pounds on my back and hiking was no longer in the cards. Deep squats and deadlifts weren't, either.

I sought mental health services to gain a foothold on my emotional well-being. There were issues to nip in the bud. Knowing I could confide in Stu, I approached him for guidance. He'd sought help, and it seemed on the surface to keep him balanced. He was a family man with a wife and kids, so he had no room for error. Stu gave me the scoop, and I scheduled an appointment with mental health. I had never spoken with a psychologist before.

The psychologist referred me to a psychiatrist, who prescribed meds following a single visit. I also completed sleep studies and other procedures to tap into my head. Some meds I tried, others I flushed. I didn't want to fall into the trap of being dependent on prescription drugs. The sleep study determined I suffered from periodic limb movements. My arms and legs twitched over four hundred times in a five-hour block of sleep. I was prescribed meds for nightmares and restless legs syndrome. I tried the nightmare meds but tossed those for limb movements after discovering they were prescribed to people with Parkinson's Disease.

Following months of treatment, the medical board process became official for me, Gus, and Stu. It was a long and demoralizing process. The powers-that-be challenged every finding and diagnosis from their own medical professionals. They'd attempt to undercut a medical examiner's opinion and belittle the injured patient. If a person wasn't careful, they'd get intimidated into signing documents that reduced their diagnoses and slashed their VA benefits. A civilian doctor who filed paperwork for a regional board at Fort Lewis summoned me to his office and mentioned my lower back injuries weren't combat-related. He attempted to get me to sign documents acknowledging it.

"My back injuries aren't combat-related?" I placed my ink pen back into my blouse.

"No, Mr. Bruce."

I leaned back in the chair and stared at the doc with my *what the fuck* face. "The Alaska Spine Center said injuries of my nature were caused by wear and tear or trauma. I've been on over a hundred documented foot patrols and was in the turret of an MRAP that flipped. I think I've got those two things covered."

The civilian doctor, wearing a cheap-looking dress shirt and tie draped with a white coat, folded his arms, sneered, and stared. An uncomfortable silent pause followed.

He then attempted to explain a combat-related injury. "Say, for instance, you're on the FOB walking to a port-a-john, and you trip and break your ankle. Yes, you'd be inside a combat zone, but that isn't considered a combat-related injury because it isn't related to combat."

Feeling insulted, I paused before taking a breath and lashing out. "I wasn't on some FOB walking to a shitter, sir. I was 'outside the wire' on a nighttime combat mission performing a BDA because we'd taken fire. The vehicle crashed during a patrol."

"Sign the document, Mr. Bruce, so I can get this to the board."

"Fuck that. I ain't signing shit." I stood and left, refusing to be intimidated.

The medical evaluation board process was filled with shenanigans. The powers attempted to force soldiers to sign away their rights to appeal before cases were adjudicated. They also slipped us papers to sign without informing us that a base attorney was available to review documents. If they tricked you into signing, things were final, and an attorney couldn't assist. It happened to several dudes. If you stood up for yourself and offered concrete evidence for your case, you were shuffled into a bottomless pit. Guys felt as if they were never going home, so they'd sign away anything for the prospect of moving on. Frustration was the name of the game. It reminded me of my dad, Big Ronny B, never seeing a dime from the VA for almost forty years. The agony of filing a claim would result in him walking away and saying, "fuck it."

The frustration of being stuck got to me after I went on leave and found my father in poor condition. The pulmonary fibrosis was taking its toll, and his decline in health was rapid. The difference in his appearance from the last time I'd seen him was like night and day. When he picked me up from the airport, I noticed a large oxygen tank on the floorboard of his passenger seat. He went nowhere without it. Sometimes he rolled it around on a small dolly if he knew he'd be away from home for a while. Simple tasks were becoming difficult for him. My active-duty contract was almost up, but I was stuck until the med board was settled. I needed to get home and stay.

I returned from leave and received a phone call from my new squad leader. He needed to see me at company headquarters. He said it was important. *Maybe he's gonna tell me the med board is done, and I'm going home.*

Sergeant Finch stood by the wooden conference room table, eyeballing me with his piercing blue eyes. He smiled, extended his hand, and congratulated me. *Fuck yeah, I'm goin' home!*

"The company is presenting you with this award," Finch said as we shook hands. "I read it, and it's well-deserved. Great job, Bruce."

I received an Army Commendation Medal with a V Device for Valor. It had to do with my actions on November 4, 2009. Inside a folder was a diploma-style certificate along with a lengthy narrative. It'd been well over a year since that day. Someone working in the training room found awards for Stu, Blackie, and me stuffed inside the bottom of a file cabinet drawer. Word was that the awards were written as Bronze Stars yet knocked down to an ARCOM. Though I was appreciative and flattered, I would've traded it for an honorable discharge and a plane ticket home.

Joey C asked to read the award when I returned to the barracks. "Jesus, man, they just hand you something like this?"

"This unit has no goddamn couth," Phil said as he stood by a bunk reading a magazine. "You'd think the battalion commander, at least, would present this in a formation."

"Proud of you, buddy," Joey C said as he patted my back.

I didn't want to stand in front of a formation with a few guys left who'd displayed courage and deserved more yet received nothing. Brandon, Ross, Gus, Joey C, and Dottie would've been inside that formation. But they'd be spectators whose bravery went unrewarded. There was plenty of gallantry in my old platoon. Blackie wouldn't have made this formation to receive his award, however. He'd been chaptered out with a less than honorable discharge, and his medal was mailed to Illinois.

In July 2011, I wrote to Congressman Don Young of Alaska. I asked for a transfer to a base in Georgia or to have my çase prioritized. I emphasized my father's terminal illness. He lived alone and had reached a point where he couldn't take care of himself. The congressman was informed that my father was a disabled Vietnam veteran suffering from a disease likely due to Agent Orange. In the eighties, Big Ronny B was awarded a small settlement from a class-action lawsuit against the makers of the defoliant. *Uncle Sam should throw me a bone.*

The army responded to Congressman Young's inquiry by denying all his requests. Furious, I read the government's response. The civilian powers-that-be said that once the medical evaluation board process started, no transfers could take place. The letter also said my case couldn't be expedited because the process had to play out for my own benefit. *What a fucking joke; fuck me and my suffering father.* There was no telling how long the case would drag on. During the initial briefing, I was informed it'd be a two-hundred-eighty-day process. That timeline proved false. There was a silver lining, though. The army stated that I would be given all available leave opportunities if it didn't interfere with the medical board process.

During the summer of 2011, the battalion stood up a Rear Detachment Company. Everyone receiving an honorable discharge soon, going through the medical board process, or getting kicked out was assigned to "Rear-D." The brigade was set to deploy to Afghanistan in December; thus, "Rear-D" was the place for those not deploying. Newly promoted Captain Troutman would serve as Rear Detachment commander, and I couldn't have been happier. Captain Troutman was one of us. He'd look after his guys.

Whenever I spoke to my father on the phone, he'd say. "You don't leave that place until your situation is handled." He'd continue, "I got fucked by the system, and look at me now. I know it's frustrating; I've been there. Who knows how you'll be when you get to my age? Fuck 'em, get yours!"

In August 2011, I was given a glimmer of hope when I signed my paperwork (DA Form 3947) to go to the regional board. After the signing, I was promised I'd be out of the army within ninety days. But things didn't go smoothly.

I received a phone call from the base VA representative. "Is this Ronald Bruce?"

"Yes."

"Um, Mr. Bruce, do you have a copy of your marine medical records?"

"Yes, sir, I do."

"Is there any way you can get them to my office?"

"Not really. They're locked in a storage unit in Atlanta."

"Um, okay," he said. "The board at Fort Lewis has misplaced your marine records."

"They lost my records?" I asked in shock.

"For now."

"Geez, so what happens now?"

"I'll call you later for further guidance." He hung up.

It took three months for the board to acknowledge they'd lost my marine medical records and move forward. But the fuckery had only begun. In December, the board at Fort Lewis rejected my psychiatric diagnosis, and the case was sent back to Fort Richardson for further review. In January 2012, I signed another DA Form 3947 to kick off another ninety-day process. A second opinion was ordered, so I visited another psychiatrist, and she diagnosed me with the same condition. My case was sent back to the board at Fort Lewis, and my psychiatric diagnosis was rejected again. Another review took place.

With no end in sight, Captain Troutman approved me for leave in February. He and the Rear-D first sergeant knew I was getting fucked. So, they'd "lose" my leave forms whenever I returned so I'd have available leave days in case the system kept screwing me. It wasn't my command nor uniformed army sticking it to me. It was the civilian powers who controlled the army. Troutman and the first sergeant advocated on my behalf because they wanted me home with my father.

While home with Big Ronny B, I drove him to a medical appointment for his lungs. Doc said there was nothing that could be done, and his condition would rapidly get worse. It was heartbreaking. He and I both knew he didn't have long to live. I'd gone through the same thing with my mother, witnessing someone I loved dying from a disease. During my last week of leave, I received a phone call from a Major Gomes. He explained he'd be evaluating me when I returned from leave. This was the third time I'd be evaluated, and Major Gomes claimed he had the final say in the matter.

Major Gomes' own words were, "I'm going to be the 'clean up' guy in this whole mess, and my decision will be final."

When I returned to Alaska, Major Gomes had studied my entire file, and he examined me a few times. As the third opinion, he was the latest doc to diagnose me with PTSD, so I signed another DA Form 3947 to set another ninety-day timeline. Despite being deemed the "cleanup guy," Major Gomes' psychiatric diagnoses were rejected by the board.

I had no fucks to give about anything. It seemed the system wanted to string me out until I got myself kicked out. Perhaps I'd lose my cool, snap, and do something stupid. I called Major Gomes to tell him his diagnosis had been rejected, and I could tell that he was also furious. He'd been given a task, and his professional opinion was shit on by the people who assigned him the task.

I learned from a paralegal on Fort Richardson that the Madigan Clinic on Fort Lewis was under an army and senate investigation for downgrading and denying PTSD diagnoses. Over forty percent of PTSD diagnoses had been rejected by this board since 2007. The board even employed a forensic psychiatry team to review cases and change or disregard diagnoses. The army cleared itself and the clinic in an internal investigation. No surprise. But the Madigan Clinic stopped utilizing a forensic psychiatry team to review diagnoses since they were the only clinic in the entire military to do so.

Armed with this new information, I submitted a letter to Congressman Lynn Westmoreland's office. He represented my home district in Georgia, and I felt he'd take more of an interest. I wasn't the only one being fucked by this crooked board. Stu and Gus had already been shafted since their PTSD diagnoses were also downgraded by the forensic psychiatry team. *If I can get the congressman to take a serious look into the matter, maybe this board will be forced to do right by its soldiers.*

Now was the time to fight the system that'd been fucking dudes for too long. I made two requests to the congressman. First, investigate why three doctors evaluated me only to be overturned by a forensic psychiatry team that never laid eyes on me. Second, force the board to adjudicate my case so I could go home and care for my terminally ill father. People assume Vietnam-era veterans were the last to get screwed by the system. It's not the case.

I spoke to my friend Eric about the situation, and he put me in touch with two ladies who worked at the Shepherd Center in Atlanta. The Shepherd Center is a private, not-for-profit hospital and one of the premier spinal cord and brain rehabilitation centers in the world. With their ties due to the center's advocacy for disability rights, employees at the Shepherd Center possessed solid connections to senators, congressmen, and even high-ranking military leaders. Tina and Bonnie contacted me, and I forwarded them all information regarding my case. I'd attended high school with Tina and graduated with Bonnie's brother, so they both knew my father and me.

The ladies worked aggressively on my behalf. Tina was a practicing psychiatrist herself. She reviewed my entire case file and was appalled at how it was being handled. Tina knew I couldn't bullshit three different psychiatrists. Psychological stuff can't be read like MRIs, but trained professionals observe you over time. Not to mention my sleep study, which concluded I twitched so much that a Parkinson's Disease and anti-nightmare medication were prescribed. That's my psychological MRI. Both Tina and

Bonnie embarked on a vigorous letter-writing campaign to every legislator the Shepherd Center had ties to. They contacted each doctor who'd evaluated me and spoke with high-ranking military officials. Tina even called a retired general whom she said was a second father.

As the wheels turned to get me home, Captain Troutman called me into his office. "Bruce, have you filed an IG complaint?" He was referring to the inspector general's office.

"No, sir," I said as I sat in front of his desk. "I haven't thought about it, sir."

"Well, let's do that." Troutman leaned forward. "I know you've written the congressman and have some Atlanta contacts working this, so let's get every available resource on it."

The first sergeant stood in the doorway with his hands on his hips and chimed in. "Yeah, let's bombard those assholes with complaints coming in from every which way, so they won't know what hit 'em."

"Roger that, first sergeant."

"You hang in there, Bruce, and be a little more patient for me." The captain's lip curled into a smile, which reassured me. "We're gonna get you home. I've gotta good feeling about it."

"Thank you, sir."

I was once again called into Captain Troutman's office. As I walked inside, the captain and first sergeant were glowing. I knew something good was coming. They couldn't hold it in.

"I dunno who your friends are in Atlanta, but I wish I knew them," the captain said. "A four-star general called the board in Washington and told them to medically retire you and get you outta Alaska ASAP."

"Congratulations, Bruce," the first sergeant said while he clutched my shoulder.

A mixture of urges to laugh and cry poured over me. A two-hundred-eighty-day process took four-hundred-eighty-days to complete. A year after my original ETS date, I was headed home.

I ended another chapter in my life with emotional hugs, handshakes, and well wishes from the few of my combat buddies who remained.

CHAPTER 18: WHEN RONNY COMES MARCHING HOME AGAIN

The old church bell will peal with joy
Hurrah! Hurrah!
To welcome home our darling boy
Hurrah! Hurrah!
The village lads and lassies say
With roses, they will strew the way
And we'll all feel gay
When Johnny comes marching home

-Patrick Gilmore (1863)

During the "honeymoon period" in Georgia, I got reacquainted with old friends, visited relatives, and did things I didn't have time for during a short leave. The best part was enjoying quality time with my father. He still required 24/7 oxygen, but he had a portable system that allowed him to get around easier.

I spent nights out catching up with ole Wildfire, Emmy, and Wayne.

Like old times, Wildfire skipped lines, burped on beta males, and chatted with Hispanic bar workers in his redneck Spanish. One evening, Wildfire was arrested. I exited a bar to check on him and heard screams of, "Don't you know who the fuck I am?" as Wildfire resisted. Several officers tossed him on the concrete face-down and cuffed him. Wildfire was no stranger to a cell and didn't give a fuck about jail, though. Like a poor man's Teflon Don, nothing stuck.

I attended concerts with Shaun and Mark at the iconic Masquerade. Suicidal Tendencies, Overkill, Testament, and Hank III rocked my balls off that summer. Danzig and Hank Jr. also toured Atlanta. I'd seen both of them five times, but it'd been almost a decade since the last time. I moshed and banged my head at the Danzig show, received a fist bump from Glenn, and sang along with a rowdy crowd at the Hank Jr. concert. Ole Uncle Hook tagged along for Hank since he dug him. But he was also a big fan of Gregg Allman, who opened the show. Marijuana smoke wafted from the joint between Hook's fingers as he sang *Ramblin' Man.* Hook's excitement couldn't be contained when Otis Redding III hit the stage and sang with Gregg. I had childhood memories of a drunken and stoned Hook singing *Sittin' on the Dock of the Bay* and *Ramblin' Man* at family gatherings. Those were his favs.

One of my old friends was a former jarhead named Oscar. We'd attended high school together but saw one another rarely. After his graduation from the University of Georgia, he had joined the marines and served as a combat correspondent. A six-month tour in Iraq was part of his credentials. Oscar joked he was the "Private Joker" in the corps, referring to a character in *Full Metal Jacket.* An unapologetic flag-waiver, he was a card-carrying member of the Buckhead Young Conservatives. He had a semi-professional career, a rented home in Atlanta, and a couple of nerdy roommates.

Oscar loved hosting parties where opinionated young liberals and conservatives discussed politics. I referred to his new friends as "square glasses," a term I'd coined as a twenty-three-year-old freshman at Kennesaw State. Square glasses don't refer to any particular political affiliation; it's for bratty know-it-alls. I coined the term at Kennesaw State since many of the gals went for the "librarian look" during the early 2000s. Beta males followed suit. People used the term "hipsters" to describe these folks, but I stuck with "square glasses." Many of them wore frames for "the look" despite not needing a prescription.

With reluctance, I attended some of Oscar's parties. They were filled with dogmatic "square glasses" discussing political issues as they drank Pabst Blue Ribbons and India Pale Ales. I flashed back to when PBR was a redneck beer and laughed, knowing it was now the thing for yuppies. Some folks envision "square glasses" as entitled millennial liberals who are politically correct. I won't disagree with that assertion, but millennial right-wingers are no less of "square glasses." They get butt-hurt as easily as liberals when you disagree with their ideas. A lot of the "square glasses" attending Oscar's parties were silver spoon-fed trust fund babies. Wildfire called the hipster dudes "sweet faces." "'Cuz they look like they have no miles on 'em."

I bit my tongue as I stood on the deck of Oscar's home and listened to a group of gung-ho square glasses talking foreign policy while sipping IPAs. They uttered the usual.

"We must destroy terrorism; fight 'em over there rather than here."

"We should've never pulled outta Iraq. We need to send more troops to do it right and wipe out the Taliban. Bomb the shit out of 'em."

After pounding highballs and toking reefer with a cool little "square glasses" dude, I tossed out. "Why ain't y'all fighting 'em over there?"

The mood shifted as if music had ceased. No one answered. I regretted throwing my two cents in because I didn't want to get into it. I'll talk politics with friends, but I don't like doing so with strangers. When it comes up in bars or other public places, I ignore it or pretend to agree with what's said. But these tough-talking "square glasses" were getting on my nerves.

One young man swigged his PBR and took the bait. "What do you suppose we do, sir?"

I chuckled when he referred to me as "sir." But I appreciated the respect shown. I spat some fire at these young "square glasses."

"How 'bout bombing them into oblivion?" another gentleman asked. He sat on a chair nibbling a burger.

"If bombing the shit outta people won wars, then Vietnam, Iraq, and Afghanistan would've been won in days or weeks." I leaned against a post and raised my highball. "It's not that simple, man."

I warmed up to the "square glasses" as they drank more PBRs and IPAs and picked my brain. Although cool and respectable, these dudes were typical chickenhawks. A chickenhawk favors war yet avoids military service like the plague.

I later found Oscar and toasted his whiskey on ice. "How can a former jarhead with a deployment under his belt be a member of the Buckhead Young Chickenhawks?"

Before Oscar could reply, I added. "That's what it should be called. The Buckhead Young Chickenhawks rather than Buckhead Young Conservatives."

"You're a hoot, Bruce," Oscar said, guzzling his whiskey.

I hadn't been home long when I received bad news from the old unit. Stu sent a text. "Hey man, have you heard about Clover?"

"No. Waz up?"

"I figured you'd want to know he's no longer with us."

"Whatchu mean?"

"He took his own life."

I had been watching television with Big Ronny B, but I left the room to hide my emotions as I read texts. Another friend had committed suicide, and it hit hard. I drank Wild Turkey 101 and smoked loud-ass reefer in honor of ole Clover. Physical pain ripped my stomach. I silently toasted my glass to our little run-in with the battalion sergeant major while high as a kite on Afghan Kush. The army washed its hands of Sergeant Clover following the deployment. He'd served in Iraq and Afghanistan, developed substance abuse issues, and the army just fed him a "big chicken dinner." Now Clover and his shattered family are a statistic.

During my first month home, Big Ronny B's condition worsened. He'd stop to take a break during the short walk from his vehicle to the front door despite the portable oxygen system. He was becoming homebound. Since we understood his time was short, Big Ronny B requested that I take him to his favorite watering hole to see his old buddies one last time. I loaded up the wheelchair and oxygen tank and rolled him through the front door to surprise his old friends. People rushed over to offer handshakes and hugs. For a couple hours, he chatted with his friends, laughed, and reminisced about trips to Biloxi casinos. During the ride home, my father's face gleamed. Though I was happy for him, I felt empty inside. I knew the million-dollar charm was living on borrowed time. The little time he had left needed to be filled with comfort rather than stress. It was tough.

As the summer moved on, the pulmonary fibrosis decimated Big Ronny B. We made the best of the situation. Hook and I, along with other visitors, often sat inside dad's bedroom and watched television, listened to music, and talked about the past. He loved reminiscing about the "old days" with his younger brother Hook. I enjoyed listening. His television was fixed on the Western Channel, and he watched reruns of *Gunsmoke*, *Bonanza*, *The Rifleman*, *Rawhide*, and numerous other Westerns. One of his favorite hobbies was scratching off lottery tickets. He'd send me to the store each day to grab fifty to one hundred dollars' worth of scratch-off tickets and to drop off the winning tickets he'd scratched the previous day. Dad dropped a small fortune on the lottery.

My sister would visit whenever she could, and dad would light up like a star. Shelly's presence was always enough to shift his mood. She was his first child and only daughter, a daddy's girl. They'd sit together and discuss events from a time beyond my memory. They talked about Nanie (grandmother) and Paw Paw (grandfather) and their simple life. They'd chat about Paw Paw's coon dogs and the beagles my father raised for hunting rabbits. My sister adored them. I'm sure being around lots of hounds led to her life of rescuing sheltered dogs. Shelly stayed strong in my father's presence, which kept him at ease. I pulled Shelly to the side to discuss plans for his eventual funeral. He'd already purchased his resting place beside my mother and requested to be cremated. Before she left, I hugged her and thanked her for making dad's day. We were close despite not seeing one another much.

Big Ronny B's condition worsened before the autumn. His home oxygen system didn't provide enough liters per minute to keep him breathing comfortably. One night, while Hook was visiting, Big Ronny B fought for breath. Against his wishes, I called 911, and he was rushed to the emergency room. After he was evaluated by his pulmonary specialist, I was summoned to meet with a doctor. A hospice representative was present; it was time to prepare for my father's death. I wasn't surprised, but the news still hit like bricks. Three months after I left the army, my father was admitted into a hospice facility.

If there were ever a proper way to go out, Brightmoor hospice provided it. The facilities were top-notch, and the staff was professional and caring. The nurses sang and prayed with patients. An older black woman, who worked in housekeeping, took to Big Ronny B. She'd enter his room and sing religious hymns, which put him at ease.

When my father entered the facility, he was in poor condition and had more bad days than good. A breathing mask covered his face and made it difficult to speak.

One morning I entered the facility, and a nurse approached me. "Would you like to pray with your father? He wants to pray with you."

"Yes, ma'am."

A circle of folks surrounded Big Ronny B, including the facility chaplain, the woman who loved singing to him, and a couple of nurses. Tears were streaming down their faces. I entered the group and joined hands. Big Ronny B's mask was removed, and he said one of the most inspirational prayers I'd heard. It was shocking as if he'd gone from a vegetative state to a professional preacher. Despite being Christian, my father wasn't a religious type. Our family didn't say the blessing before meals. But at this moment, he lit up the room.

Big Ronny B's time was not quite up, though, and he snapped out of his funk. He carried on conversations as if he were in the prime of his life. By no means was he returning to health, but he'd leave this world with more dignity.

On Veteran's Day, a uniformed military NCO walked the hallways, visited each patient, and presented veterans with a citation. Big Ronny B's read. "This certificate is presented to Ronny Bruce in recognition for service in the United States Army, Vietnam. Thank you for your courage and commitment. Your service will never be forgotten." *The old 'Nam Vet finally got his thanks.*

Big Ronny B was outliving his prognosis. A social worker asked if I had the means to take him back and care for him with the addition of home hospice resources. Because I wasn't a nurse, I was taken aback. But it was either this or send him to a nursing home. My father wouldn't be comfortable living his final days inside a nursing home, and I'd feel terrible sending him there. A professional caretaker presented herself during the eleventh hour, which was a godsend. Connie had been watching over terminally ill people for years and happened to be the sister of one of my father's childhood friends. She even attended high school with Big Ronny B. Connie cared for my father inside his home.

During the fall and winter of 2012, Big Ronny B and I watched football and took in visitors. He continued scratching lottery tickets. Living with a terminally ill person is a rollercoaster ride. There are mood swings, moments of limited awareness, and good days mixed with bad. On some bad days, I'd call my sister to ask for advice. She helped me deal with the tough situation.

Sometimes when he was delusional, dad would say. "Son, we're gonna go to some Georgia Tech games and get Falcons' season tickets next season."

Choked up, I'd reply. "Yeah dad, that'd be awesome."

Other times, Big Ronny B kept it real when I had my own lapses in judgment. "You don't need all that sugar in your coffee, dad."

"But I'm dying, son."

Depression zapped Big Ronny B during bad days. He despised being bed-ridden. Though he was once a hard-working man, now he required Connie and the hospice staff to bathe him. He had conversations with himself during episodes in the middle of the night. He played his own part as well as the part of the people he spoke with. Since I had my own issues with sleep, I entered my father's room and lashed out, demanding to know who he was speaking to.

He snapped awake and said. "What the hell are you talking about?"

When I left the room, he faded back into these conversations. The names Steck and Charlie were often mentioned, and I knew they were two of his best friends while in Vietnam. He also spoke with his deceased brothers Larry and Kenny. It went on all night.

One night he screamed to Steck and Charlie, "Just shoot the motherfucker," as he tossed in his sleep.

I called Shelly. "Dad's throwing up at night and screaming in his sleep. I dunno what to do."

"I know it's hard right now but hang in there," Shelly said over the phone. "Imagine how tough it must be when you're gasping for air and can barely breathe, even with oxygen. Enjoy the good days because he doesn't have much longer."

I thanked my sister for her pragmatic words, and we exchanged goodbyes. The next morning Connie arrived, and I told her my father had been sick and spent the entire evening talking to friends and fam from the past.

Connie stood across from me and placed her hand on my shoulder. "A lotta times when people are about to die, they talk to the dead. I don't think he has much longer, Little Ronny."

I had too much on my plate to give a fuck, but election season was going full bore during the fall of 2012. As *South Park* put it, the election pitted a giant douche versus a turd sandwich, anyway. I didn't vote. I don't do the "lesser of two evils" thing because I feel like my vote is an endorsement. Despite hearing compelling reasons to vote from both sides of the spectrum, I wasn't persuaded. Neither candidate met the threshold to deserve my vote. The Obama Administration didn't give me a warm fuzzy after what'd been going on, and I wasn't down with Republican Mitt Romney, either. Mitt had received several draft deferments to avoid the Vietnam War. Why would I rush to the polls and vote for that guy? Why would I stamp my approval on a candidate who dodged a war while my father was dying an agonizing death from an illness he contracted during the same war?

The only good thing about the 2012 election, according to Big Ronny B, was that his idol Clint Eastwood spoke at the Republican National Convention.

Despite the rollercoaster rides of the election and his illness, there were more good days than bad. Connie and I made it a point to be positive and help my father live his last days in peace. She sat in his room each day. They took in gun-slinging westerns and laughed with *Judge Judy's* style of justice. On weekends, Hook and I watched football with my father. There was no shortage of visitors, either. They were often my own friends, and they turned out because they looked up to Big Ronny B.

On a routine Thursday morning, Connie arrived for her shift. She brought balloons and flowers because it was Valentine's Day. Big Ronny B appeared in good spirits, and he drank his morning coffee and laughed through a *Sanford and Son* episode.

"I'm gonna head to the gym and get a workout, dad." I stood by his bed and admired the balloons and flowers. "That's sweet of you, Connie."

"Okey-doke, son. Get one for me too," Big Ronny B said as he raised his arms and flexed his muscles.

I spoke to Connie for a minute on my way out and drove to the gym. I wasn't inside the gym long when a young lady who worked there approached me as I spoke to a group of Cross Fitters. A tear smeared her cheek.

She inhaled a couple slow breaths before she spoke, and I knew what time it was. "Ronny."

"Yes, ma'am."

"Someone called and said to get you because your father passed away."

I stood motionless by a group of people who didn't know my situation. *She could've pulled me to the side and done this in private.* A young CrossFit gal started weeping as another gentleman patted me on the shoulder.

I turned and faced everyone. "It's okay; my father was terminally ill."

Overcome with emotion, I couldn't help but think about the night before. My father had spoken to Steck, Charlie, and his deceased brothers and sisters.

I entered our home, and Connie was waiting at the door. The hospice nurse stepped out of Big Ronny B's room, sobbing.

"It's okay, Little Ronny." Connie clutched me in a hug. "He's finally going home to be with your mother and the Lord."

I couldn't speak.

"Would you like to see him?"

I nodded. With flowers sitting atop a table and balloons floating in the air, I wept over my father's lifeless body. He appeared at peace. His suffering was over. I stood by his side and told him I loved him, over and over. What an appropriate time for him to reunite with my mother: Valentine's Day.

The next day, I hung with Uncle Hook. I wanted to reminisce about ole Big Ronny B. I drove Hook to a nearby tobacco shop so he could buy cigs. I followed him inside the store with no purpose in mind. Recognizing the lottery section, I bought four five-dollar scratch-off tickets on a whim. They had a picture of President Grant and a fifty-dollar bill on the front. I never played the lottery. *Do it for Big Ronny B.*

Hook and I returned to his home, and I said, "Scratch these tickets while I go use the bathroom."

Hook commenced scratching with his lucky coin. "Oh my God! Little Ronny, you gotta come here. Hurry!"

"What?"

"You won twenty thousand fuckin' dollars!"

"Get the fuck outta here."

While studying the ticket, I almost hit the floor. I'd never won a substantial prize in my life, never been lucky. I smiled and looked upwards. For a moment, I felt at peace.

Big Ronny B was sent off from this world right. His service occurred at his cousin's funeral home, and his ashes were placed inside a mausoleum beside my mother's. Many gathered to pay tribute to a man who led a life well-lived. We viewed a DVD, set to music, filled with pictures from my father's life. The preacher eulogized Big Ronny B based on the obituary I'd written for the Atlanta Journal and Constitution. He mentioned my father's love for the Georgia Tech Yellow Jackets. Hook let out a semi-drunken "boo" since he was a Georgia Bulldog; everyone got a quick laugh.

As the family prepared to exit the chapel, Elvis sang through the speakers and brought more tears to our eyes. At the base of the angelic Cleveland mountains, a classy entombment ceremony befitting a warrior took place. Decked out in dress blue uniforms, the NCOs looked professional as military honors were bestowed upon Big Ronny B. The civilian gathering was impressed. The folding of the flag was precise, and the bugler's rendition of "Taps" was on point, eliciting more tears. The soldiers rendered a slow salute to Big Ronny B's ashes as they sat in their final resting place.

The sharp-dressed NCO hugged the folded American Flag and knelt to my sister's front. "Ma'am, I present this flag to you on behalf of a grateful nation."

He handed over the flag, popped to attention, and gave a slow hand salute one final time. My sister passed me the flag after the ceremony, and I held that flag tight all afternoon. My mother and father were together again.

CHAPTER 19: THE WOLF

Over the next several years, I went into hiding and avoided the world, steering clear of friends, family, and acquaintances. It was almost impossible to rise out of the funk that buried me. If a nagging friend blew up my phone with calls and texts, I'd come up with a lame excuse for why I couldn't come out and play. One of the worst feelings was to hear my phone buzz. I wanted no part of any proposed activity.

I'd fallen victim to the ways of the wolf, who lives on the fringe of society. One can liken the wolf's existence to the rings surrounding Saturn. They stretch thousands of miles from the planet's equator like how the wolf roams far from what's taking place in society. Like the wolf, I dwelled over a hundred thousand miles from everyday society. The wolf is the ultimate recluse. Misery and pain consume the wolf, but this kind of misery doesn't love company. For the most part, the wolf will wallow in his own agony because he's content with these feelings. Yet he'd never wish his torment upon his worst enemy. The wolf will ride this wave and torture himself until he loses everything. A marked trait of the wolf is not feeling worthy of being in others' company because he feels inferior and not deserving of people's time. I was a wolf for years.

During this self-imposed exile, I also tried to grab the world by the balls. My resume was strong and displayed a wide array of skillsets. I attended job fairs, passed out resumes, maintained a healthy fitness level, and dressed immaculately. Exploring new opportunities excited me—for a minute. I'd done a good job of saving money from my military years. Winning the lottery didn't hurt, either. Yet my efforts to gain suitable employment fell short. Callbacks were few and far between, and interviews were seldom scheduled. Whenever I felt I'd knocked an interview out of the park, I'd receive a "sorry, but we've decided to pursue other candidates." Unemployment and underemployment became my reality.

Looking back, the energy I put out to potential employers was negative. One cannot live like the wolf and release positive energy. A phony smile doesn't hide misery. Recurring nightmares sometimes woke me up sweating, shaking, heart pounding. When I peeled back the sweat-soaked sheets and peeked at the alarm, I'd realize I needed to be at work in less than an hour. What kind of energy do you think I was letting off at my shitty job? During my twenties, I charmed my way into positions I barely qualified for. As a fresh college graduate, I was chosen for a teaching and coaching position over ten more experienced candidates. It helped that my former wrestling coach was the principal, but he made it known I had to earn the job.

I possessed no such magic anymore. I wasn't the only veteran struggling with employment, though. The unemployment rate for Iraq and Afghanistan veterans was over triple the national average. Either we're viewed as damaged goods or we exert bad energy, or both.

In the spring, I aggravated my lower back injuries to the point of being in tremendous discomfort. I can take a ton of pain, but I couldn't limp five steps without needing to squat. Lying in bed was my only relief. So, I made an appointment with the VA pain clinic. Beet red with pain, I entered the overfilled Atlanta VA Medical Center and was herded along with the crowd.

Thank God I'd arrived an hour early since it took that long to find a parking spot. While checking in with the receptionist, I heard a conversation from the waiting room around the corner. One guy was telling another he'd been stationed in Alaska and served in Iraq and Afghanistan.

With my curiosity piqued, I entered the waiting room and asked, "Who was stationed in Alaska?"

The room full of African American gentlemen looked up, and one of the men said, "That's me. I was at Fort Richardson."

"Oh yeah? I was also at Fort Richardson," I said. "You must've been in the 4/25."

"Yep, I was in the Cav."

"No shit? I was in the 509th."

"My man, what's goin' on, bro?" The man offered his right hand to shake.

I reached out and noticed he was missing his left arm. "Ah, you know, man, tryin' to get a lil help around here."

We laughed.

I took the seat next to him, and we talked about Alaska, Afghanistan, Bowe Bergdahl, and what we were doing now. He told the story behind his arm getting blown off. He'd been the TC of a Humvee that caught an RPG through the windshield. His left arm was severed, and his gunner lost his right leg. The gentleman was an indirect victim of "stop-loss", which infuriated me.

"Yeah, man, I had orders to Korea because I was given a choice for reenlisting, and Korea is a non-deploying duty station." The brother slumped further into his chair. "Dawg, I'd done fifteen months in Iraq and wanted to go to Korea to catch a break. But, when stop-loss orders came from the brigade commander, my shit got canked, and I was forced to go on that Afghanistan deployment with y'all."

"Motherfucker," I said as I leaned back in my chair.

"Yep, I'd still be in the army but instead, I'm sittin' here like this."

"Well, God bless you, brother." I offered my fist for a fist bump. "Hope everything works out from now on."

"I appreciated it, bro."

Despite the young man's obvious disability, he appeared in good spirits. His attitude was positive, and he had a good sense of humor. Our conversation was relaxing, and this young brother served as an inspiration. I had no reason to feel sorry for myself.

Despite taking steps forward, the hits kept coming. When I made progress, the VA took actions that'd send me into a tailspin. I felt my country had betrayed me and spit on my father's legacy. While Big Ronny B was in hospice care, he made me a co-account holder on his banking account so I could take care of biz following his death. He'd been approved for a VA disability claim two months before he died. The official VA letter stated his claim was adjudicated in December 2012 and backdated to July 2006. He passed away in February 2013. Over thirty thousand dollars, which served as backpay, was deposited into the joint banking account. A week after my father passed, I drove to the Atlanta Regional VA office to stop his disability compensation. While at the office, I explained that over thirty thousand dollars was deposited into his account. The employee viewed all documents.

"Sir, this money belongs to your father's estate since it was approved before his death." The woman stood and walked to the other end of the office. "But let me go speak to my supervisor."

She returned from her supervisor's office, smiled, and said. "Yes, sir, I double-checked with my supervisor, and he said all VA payments belong to the estate since everything was approved before your father passed."

After receiving the official word from the VA, I wired my sister half the money because I didn't want to be greedy. But, six months after my father passed, the VA confiscated the money from the account, leaving it twenty-thousand dollars in the red. I was never sent a letter informing me of the confiscation, and no court order was issued. I found out about this fiasco when the card for that account was declined when I attempted to make a purchase. I called the bank to figure out what happened.

The representative on the line refused to give any information regarding the account. "Ma'am, am I gonna be responsible for this negative balance?" I asked.

"You sure will."

"Then I need to know who confiscated the money."

"Sorry, I'm not at liberty to discuss that info."

"How in the FUCK can you not give me information about my own account when you're telling me I owe a twenty-thousand-dollar negative balance?"

She hung up.

Through a friend, I was connected with a retired naval officer living in Washington D.C. His wife was an attorney working for one of the three-letter government agencies. Doctor Bob, as he was known, was a member of the same bank. He was also an advocate for veterans. The first thing Doctor Bob did was contact the bank to get to the bottom of the situation. Since he flexed his influence, the bank admitted that the VA had confiscated the money. The VA didn't give a reason, and there was nothing the bank could do about it. I learned that no account in this country is safe from the federal government's hands. If they want your money, they can get it. The bank did apologize for how their representative treated me. Still, they also mentioned if I didn't repay the negative balance, it'd be handed over to collections. My credit and father's good name would be destroyed.

Doctor Bob turned his attention to the VA. He called and wrote letters to every official he could think of to right this wrong. I wrote Congressman

Westmoreland's office again and included all the relevant documents. I felt confident about the outcome because all the official VA documents substantiated my claims. It was clear that my father's claim was approved before he died.

Despite the evidence, the VA left a response with Congressman Westmoreland's office stating their actions were legal and appropriate. I felt I had no voice, nor had I been given a chance to defend myself. Where was my due process? Where was my protection from an unlawful seizure of personal property? Had the VA sued me, I'd be given a chance to present my documents to a court and argue my case. The VA could state its case, and an arbiter would decide based on the evidence. Is that not the way it's supposed to work in this country?

After everything my father and I'd done for this nation, what did we receive as thanks? A green weenie and a fuck you. Large banks were receiving government bailouts to the tune of hundreds of billions of dollars. High-level VA employees were receiving millions in bonuses. But the federal government plucked thirty-seven thousand dollars from a combat veteran.

After the VA and banking ordeal, Oscar and I attended a Braves game. I made sure to be taking a piss when the National Anthem played. There was no way I could stand with my hat in my hand during a touchy-feely moment of phony patriotism. Days later, I gathered the flag presented at my father's funeral, along with the folded flag mailed to me for my medical retirement and drove to Uncle Hook's house.

I knocked, Hook answered, and I said. "I need to use your burn pit."

"Whatchu need to burn?" Hook stood in the doorway puffing a Winston and gripping a tallboy Busch.

I raised the two flags.

"I don't blame you," Hook said. "After how they've treated you, fuck 'em."

I tossed tree limbs into the pit, doused it with lighter fluid, and torched that bitch. The flames stoked nice and high. I sat by the fire with a Busch in hand, and both flags clutched to my chest. Tears streamed down my cheeks, and my mind raced. I knew what I wanted to do, what I needed to do. I gripped the flags and thought about my time in both branches of the military. I remembered the guys, my father, and the men who were no longer with us. I couldn't bring myself to do it, though. I smothered the flames and walked into Hook's home, demoralized. I couldn't be like some piece of shit pissing on a flag during a campus protest. Instead of, *fuck this country,* I readjusted my tune to *fuck this government.*

At the time, the VA had been rocked by several scandals. They included the lengthy claims process, veterans dying while awaiting care, and other disgraceful practices. One night, while toking a lefthanded cig and flipping channels, I happened upon a popular leftwing pundit blasting the VA. I'd seen this woman report from Afghanistan following my tour and found her stories honest and fair. I tuned in several nights in a row to view her segment exposing the VA fraud. She was vigorous in her fight for the rights of veterans. If a leftwing lesbian female leads a fight versus the corrupt VA, you go, girl.

In addition to the national news, the local Atlanta news presented stories about the mistreatment of veterans in Georgia. Friends and family urged me to go to the media with my case, but I only made a half-assed effort. If a local investigative reporter read my documents, they'd be shocked and want to put it on camera. But I wasn't into it. I wanted to be as far from a camera as possible. There's no doubt the media could've helped but getting fucked over and taking it seemed more appealing than fighting the system, as I'd done before.

Instead of flourishing after my military service, I continued down the road of self-destruction. I stared at four walls, self-medicated, and drank more than I should've. I worked unfulfilling jobs and further secluded myself from the world. I wasn't stylin' and profilin' anymore. The ladies were nonexistent, and I didn't seek them out. Young sweet Kathy had moved on while I was getting jerked around in Alaska. Plus, my fit body had softened since I'd stopped working out. My health was slipping, and I drank to mask my depression.

When I'm at my best, I may have a drink over the weekend and a toke of reefer here and there. However, while being the wolf consumed my life, I drank more.

I had to learn how to become a part of society again, but it was tough to fight so many battles. If alcohol abuse is the cause for a person's downfall, that person can focus one hundred percent on alcoholism. If the habit is kicked, a healthy lifestyle may await. It's by no means easy. I've witnessed folks lose everything on account of alcohol and drug abuse. But when you find yourself at a point where you can't partake in everyday society for unclear reasons, where do you begin in your recovery? Where should your focus be?

I continued to dwell in the cellar. Matters weren't helped when I awoke one morning and noticed several texts from guys from the old platoon.

"Paris is dead," they read.

I heard Paris had taken his own life, another debilitating blow. Left with mixed feelings of anger, sadness, and pain, I wept over another brother deciding to take the ultimate step. I never realized that Paris was depressed. He was a happy-go-lucky type. But I hadn't seen or spoken to Paris since he left the army. Serious depression often doesn't manifest in soldiers until they leave the military – when they attempt to fit into a way of life that may have left them behind.

I stored away fond memories of Paris, like listening to everything from 8Ball and MJG to Skynyrd with him. My favorite quote of his was, "it's an addiction," when he spoke about combat. Paris was another casualty of the army's cutbacks. They busted him in rank and kicked him out for a single incident while serving in Second Platoon. He redeemed himself, in my opinion, with Third Platoon. But good ole Uncle Sam allowed him to complete the deployment, only to feed him "a big chicken dinner" once the unit arrived in Alaska. I'd heard Paris joined the reserves when he made it back to Memphis, but I guess the reserves didn't feed his addiction.

Over five years after the Afghan tour, Stan sent me a video of one of our firefights that'd been posted to YouTube. I watched it over and over. We may not have been filming, but the enemy's cameras were rolling. The five-minute clip was an edited Taliban propaganda video. Complete with subtitles and commentary, it showed the ambush inside the Margah wadi where KG was seriously wounded.

The video opened with pictures of a guerilla fighter and an MRAP pitted against one another. It was captioned—Mujahideen versus US Cougar. Rounds impacted Hix's vehicle as the MRAP I sat in moved into view. The scunion Joey C laid down from the .50 was shown for a brief period. But the video was edited, focusing on our initial shock while leaving out our clap-back. We appear to get slaughtered. RPGs exploding into Hix's vehicle were shown several times in slow motion as the words at the bottom of the screen described the decimation of Americans. While filming from the opposite side of the wadi, the cameraman shouted, "Allah Akbar, Allah Akbar," in a fevered pitch. Following the edited victory, the video ended with the helicopter evacuating our four casualties while words scrolled by, proclaiming they'd killed four Americans.

Four Americans weren't killed, but the damage was done. The video cut deep. My war caught on film happened to be a chopped-up Haji video with a Jihadi videographer chanting "Allah Akbar" during the madness. This was the mission we'd undertaken while looking for Bergdahl. Had he not taken off into the mountains, we wouldn't have been sitting in that wadi. Anger from all sides smacked me upside the head. YouTube removed the video days later for violating their standards.

My downward spiral crashed. Increased depression is a bitch too. You dial up every failure in life and place them front and center. Despite these failures being irrelevant at the time, you prioritize them. Hell, they may be over a decade old, but they still consume you as if they're fresh. Imagine doing this while already defeated? When I'd determined that I was dead to the core, the wolf sat whiskey drunk with a loaded pistol by his side.

CHAPTER 20: THE COST OF BERGDAHL'S FREEDOM

One spring weekend morning in 2014, I brewed a pot of coffee and turned on the television. As I sipped a Colombian roast, a special report announced Bergdahl's release from captivity. He'd been under the thumb of the Haqqani Network for five years. A*t least they got the guy.*

A few hours later, a ceremony was held inside the White House Rose Garden. President Obama welcomed Bergdahl's parents to announce the return of what looked to be an American hero. "Today, families across America share in the joy that I know you all feel."

He proceeded to hug Mr. and Mrs. Bergdahl and added a smooch on the cheek of the missus. My stomach ached, and my mind twisted. I popped a bottle of Wild Turkey to ease my anger and disbelief. My old platoon buddies texted one another, and they, too, felt the burn. This charade helped fuel my tormented 2014.

Questions stormed my head: Why would the US Government put on a dog-and-pony show inside the Rose Garden to welcome home a deserter? Did they realize people were injured and killed searching for this DUSTWUN dude? Did the administration miscalculate? Did they bank on this being the president's 'brokering a deal to bring our American hero POW home' moment? The president knows the circumstances of Bergdahl's disappearance, right? If I knew as a grunt, the president had to know too.

I don't hate President Obama and question his legitimacy as the president like some in our country. He ran his race and earned his victory. I don't view one side of the political spectrum as evil and the other side as righteous. Both sides have done their share of both improving and fucking this country.

My refusal to blindly support a political party prevents me from painting myself into a corner as a Republican or Democrat. I'm American. Some folks' political party affiliation supersedes their American citizenship. These peeps despise fellow Americans from the opposition party more than our nation's actual enemies. If I choose to participate on election day, I'll research the man or woman running. But most cycles, I stay home. Although I'm not a fan of modern politicians, I'll give credit when credit is due.

So, when I say the Obama administration's handling of the Bowe Bergdahl situation leaves a foul taste in my mouth, I say it with no political ax to grind.

The truth seeped out as ticker-tape parades were planned in Bergdahl's hometown. Those in the know took to social media to express their anger and frustration. Members of Bergdahl's former platoon spoke out. The tide turned. Americans were receiving the real scoop, and pressure mounted for Bergdahl to be held accountable for his actions. Many soldiers sacrificed a ton to ensure his safe recovery.

Some folks take issue with the swap of the "Taliban Five" from Guantanamo Bay for Bergdahl. "The US never negotiates with terrorists!"

That's nothing but rhetoric. The recent administrations have negotiated with the bad guys to recover people we'll never know existed. President Obama isn't the first and won't be the last to make these deals. Did President Reagan not negotiate with terrorists to secure the release of several American hostages held in Lebanon by Iranian terrorists? Iran-Contra was a big deal. The US and other Western nations always get duped in these swaps; we give more and receive less. We value life, exhausting options to get our people back, even if those people wronged us. Bergdahl deserved a passage home and a day in court. Again, I have no problem with the trade. My gripe has to do with leading the American people to believe a hero was coming home.

As a result of speaking out, the young men who served with Bergdahl entered the crosshairs of people whose sole purpose was to defend politicians. These pundits, with their money, power, and platforms, go to great lengths to keep a public official looking squeaky clean. They savaged Bergdahl's platoon mates and their stories. Most went straight for the jugular of the unit itself. During normal times, these attacks are reserved for other politicians who've agreed to make politics a part of their lives, not grunts.

Brandon Friedman, the former Director of Digital Media at the US Department of Veterans Affairs, asked on a public forum. "What if the platoon was long on psychopaths and short on leadership?"

Friedman also accused Bergdahl's platoon mates of "swift-boating" Bergdahl. What is swift-boating? During the 2004 election cycle, a group of swift boat veterans attacked the credibility of Democratic presidential candidate John Kerry's service in Vietnam. It should be noted that Kerry was awarded the Silver Star, Bronze Star for Valor, and three Purple Hearts. Whether someone agrees with Kerry's politics or not, his service record speaks for itself. As a result of these attacks on Kerry's service, the term swift-boating came to be known as an unfair, unjust, or untrue political attack. So, according to Friedman, Bergdahl was swift-boated by members of his former platoon.

To add more salt to the wound, Deputy Spokesperson of the State Department Marie Harf was told by a reporter. "Well, I think his (Bergdahl's) squad mates have the best indication of what was happening that night."

In a dismissive tone, Harf clapped back. "I don't think that's the case."

The administration was shitting on soldiers to defend a deserter. Bergdahl's platoon mates and the unit had been thrown under the bus.

Allow me to set the record straight for the Spartan Brigade, the 501st, and the men of Bergdahl's platoon. For the dudes who've been called psychopaths, swift-boaters, and liars. The NCOs in Bergdahl's platoon were experienced combat veterans. Most had served an extended fifteen-month tour during the Iraq Surge of 2006-2007, the deadliest point in the war. Bergdahl's unit was awarded the Valorous Unit Award for its successful deployment to Iraq – the second highest unit award given for combat operations. That isn't a participation trophy.

Leaders in the 501st weren't psychopaths, swift-boaters, and liars. Bergdahl's company first sergeant happened to be Third Platoon's platoon sergeant before Manny Fresh took charge. He led guys like Stan, Frankie, and The Duke in Iraq; I never heard them complain about their former boss. In fact, they appeared proud to have served under him. Yet, he was relieved of his duties as company first sergeant over the Bergdahl fiasco.

Would you believe Brandon Friedman is an infantry combat veteran? He was an officer, to be exact, who served in Iraq and Afghanistan. It boggles my mind that Friedman spit venom at the unit Bergdahl served. He should've invested more time researching before tossing grenades at fellow veterans. But Friedman would rather protect a politician's reputation, even to the point of throwing shade at fellow soldiers. He epitomizes the military slang term "blue falcon." With his high and mighty bureaucratic position in government, Brandon Friedman is a buddy fucker. He forgot where he came from.

In 2017, Bergdahl pled guilty to desertion and misbehavior before the enemy. His punishment included a dishonorable discharge but no jail time. Tons of people were furious over the sentence. Oldtimers barked – "he'd be shot in my day!" But, after all the anger, I'd made my peace. I didn't believe any judge would incarcerate him after he'd spent five years getting raped and tortured inside a Taliban hell. I predicted he'd get time already served, and I moved on before a judge hit the gavel.

Following the sentencing, President Trump tweeted. "The decision on Sergeant Bergdahl is a complete and total disgrace to our Country and to our Military."

CHAPTER 21: REDEMPTION

After dwelling inside the darkness for far too long, I knew I had to jettison the wolf's ways. The seclusion and depression were taking a toll. I wanted the old Ronny B back. But wanting something and getting it are two different things. I didn't know how to get my life together. The physically fit, happy, productive Ronny B was what I aspired to be. It wouldn't occur overnight. I had to undo negativities that'd been programmed into me over the last few years. I needed to confront my demons and let go of the anger and the frustration over the times I'd been fucked by the system. The grief I felt over my parents' deaths needed to ease. They'd moved on and would tell me to do the same if they could. I had to come to grips with the legacy of my service.

It took a while, but I discovered a road to wellness. My journey isn't for everyone, nor are my methods for achieving balance. But there's a segment of the population who might benefit from the techniques. Many are like me. This is a roadmap for a wolf desiring to ditch depression and chase happiness. I lay it out because someone who's at their wit's end might see it and decide to live and discover a purpose.

Physical health became my priority. It's one thing I have one hundred percent control over. I don't have complete control over how others view me, how much the market is willing to pay for my services, or where I fit into society. But I'm the master over what I put inside my mouth and how much exercise I get. My mental health is tied to my physical well-being. I cannot live fat and happy. When I was overweight, I felt sluggish and bloated. My blood pressure and cholesterol were elevated, and my overall numbers spiked. My back pain was excruciating and limited my mobility. My anxiety rose with the elevated BP, and regular breathing felt like a chore. I was miserable. When my physical health is fire, there's no stopping me. My mental well-being thrives. People perceive me positively because I give off better energy. Good energies create more positive relationships, which is key to navigating society.

Describing specific workouts and explaining what foods to eat isn't the purpose here. But one must exercise and eat well to achieve and maintain good health. People must discover what they're capable of and push that boundary to get their hearts pumping. I designed workouts that wouldn't aggravate my past injuries yet also challenged me. Weightlifting and cardio balanced me out. I needed to walk the tightrope between preserving fitness and limiting wear-and-tear on my body. I'm not a fan of redlining, so I move at three-quarter speed. A sedentary person would consider my workouts brutal. Prime athletes would say they're a breeze.

I learned – don't be beholden to an indoor gym. Take workouts outside. The sun, clouds, cold weather, and vistas improve physical and mental health. Grinding through a level of discomfort makes a man . . . or a woman out of you. Jog or hike your favorite trails through parks and cemeteries. Remove the buds from your ears and listen to the sounds of nature. We get enough artificial noise piped into our heads on the reg.

There's no reason to diet and starve yourself. Rather, eat plenty of nutritious foods, and drink the fuck out of water. My recipe for success was eliminating most sugar and processed foods. Being a southerner, I grew up on fried foods and sweet tea filled with a brick of sugar. I processed it well as a young man, but we eventually pay for the bad food choices we make. Baked salmon, chicken, steak, and tons of greens now fill my plate. My food choices are boring and routine, yet I'm used to them; I feel guilty if I steer away from my norm. I feel great eating boring foods versus stuffing my face with burgers and fries. It's more than worth it. I'm not a food snob, though. When I attend a cookout, I'll eat three cheeseburgers, munch on chips, and drink heavy beers with the fam or friends. But you can bet your ass that for at least a week before the event, I've eaten clean and exercised a lot. After a blowout, I'll fast for hours, and my next workout will have a little extra sauce on it. It's a mindfuck but worth it.

Getting my health under control provided benefits to all aspects of my life. My pain and inflammation were reduced, my anxiety decreased, the headaches went away, and my nightmares occurred less and less. I was a new man. And the illegal supplements I'd experimented with before? They aren't necessary when you're doing things right. If anything, they elevate BP and damage the liver. It defeats the purpose of living well. The purpose isn't only to look good, but also to feel good. Oh, and the cigarette-smoking I did on deployment? I had no problem dumping the habit since I never was a full-time smoker outside the military.

I sometimes read about high school acquaintances dying from illnesses brought on by an unhealthy lifestyle. Bad health bites us in the ass in our forties. When someone dies of an obesity-related issue at forty, I can't imagine how weak and sluggish that person must've felt for years. It doesn't have to be that way. If I drop dead tomorrow from a heart attack, despite living healthier, then know I felt wonderful for years leading up to it.

This new regimen allowed me to flush the pain meds and antidepressants. My daily pill intake was limited to a cocktail of over-the-counter vitamins.

Through the years, I'd been prescribed more than a dozen medicines. I wanted no part of them anymore. On meds, my anxiety lessened, but the tradeoff was moping around with no urgency or purpose. The side effects include suicidal thoughts, and I'd been down that road. I wanted a clean, non-medicated, natural state of mind.

A relocation was also crucial for my continued growth. A troubled person cannot reinvent themselves inside the same town with all their baggage. As I'd done before, I fled Atlanta and moved to another state. Fresh grounds, people, and perspectives were needed. For someone else, Atlanta is new and fresh; for me, it's where my demons live. Moving was easy; being a wolf has a few advantages. I'm a man of the world. Since leaving the military, I've moved five times and held a driver's license in three states. It's intimidating for some but welcomed by a ramblin' man. Gaining knowledge through new experiences is a value of mine. It drives me and keeps me going.

Some folks are grown, yet they've never lived outside the zip they were raised in. They've never attended college outside of town, left for the military, or accepted a job on the other side of the city. The only time they've experienced a new scene was while on vacation. Their world is small, and perspectives limited. They live inside an echo chamber among like-minded individuals with similar beliefs. It's fine for people who dig it, but that life isn't for me.

Now that I was rollin' and gathering momentum, it was time to remove toxic people from my life. I'd fled the baggage, so why not "ghost" haters and toss the trash? Someone who shares your last name or DNA can be dropped if they do nothing but shit on you. Yes, I give my family the benefit of the doubt. But if the family member is toxic, I cut ties. I dropped a few family members and past friends because they were nothing but bad news. Once the haters are gone, you can surround yourself with positive and encouraging people, who support your goals and dreams and never diss you for trying to better yourself. They'll encourage you to go farther, shoot for the moon. Make no mistake, these same people will call you out if you're slipping. They're honest and real. Haters are miserable and unsatisfied with life. Drop them like a bad habit and never look back.

Once life is moving in a solid direction, the easiest step to wellness comes into play. It's a simple concept. Be kind to others and perform good deeds. Use that brewing positive energy and release it to others. Our smile, wave, or display of empathy might be the only form of humanity that a struggling person witnesses all day— or month. I believe in karma; what goes around comes around. But we shouldn't force ourselves to be kind for rewards. Years later, you'll look around and notice the people surrounding you are the supportive ones. Good energy begets good energy.

Be kind everywhere. Tip servers and other service workers more than the norm. If you can't afford to tip well, don't use the service. Their livelihoods are dependent upon gratuity. When businesses are shutting down due to a pandemic, pull a c-note from your money clip and pass it on. You may be the last customer for a hot minute. Hand the barista a ten-dollar bill after receiving your morning fix and say, "Keep the change, ma'am," if you can. Spread the cheddar to those essential workers if your finances are in order.

I once read a social media posting from an acquaintance who described how he dealt with restaurant servers. To make a long posting short, he enters the joint with a twenty-percent tip in mind. Then he makes deductions based off whatever inadequacies he experiences. The only thing I deduced from the post is what a miserable son of a bitch the dude must be. He wants to be served by peasants. Don't be that dude.

For the love of God, don't post your good deeds on social media for the public to admire. It goes back to not expecting something in return. If you crave more gratification than the good deed provided, reevaluate yourself. Let me clarify; posting pics from a charity event that you attended or from raising money at a Relay for Life walk is positive. You're among friends and having a good time doing something for a noble cause. But don't share to social media how you jumped a stranger's dead battery or handed some cash to a homeless man on the corner. It erodes the pureness. If anything gets posted, let it come from the stranger whose battery got a boost or the homeless man who was blessed with money that day.

These techniques came to fruition for me while I was living in Colorado, breathing the crisp mountain air of the Rockies. Demons fizzled in my rearview while exciting opportunities awaited. Rather than patrolling opium fields in Afghanistan, I was standing guard over acres and miles of marijuana crops. The scope of the legal marijuana industry is mind-blowing, a story in itself. One site was located four miles down a dirt road. It included two multimillion-dollar greenhouses the size of football fields. Outside, plants aligned in perfect rows stretching for acres – a marijuana jungle. It had top-flight security systems: gated entries, cameras, motion detectors, and alarms. During a night shift, the midnight moon illuminated a mountainous backdrop.

When asked if we had any questions during the job orientation, one young man raised his hand. "Since we're contractors, do we get employee discounts?"

"Sure." The former jarhead Iraq vet stroked his beard. "Leave your firearm behind, and don't wear your company shirt inside. Cool?"

On my first shift, a black dude who was a former medic in Iraq asked. "You smoke?"

"I'm down, man."

"Shit yeah, dawg." The brother sparked a blunt with his army zippo.

We patrolled the marijuana jungle, passing a blunt back-and-forth, with guns resting at the low ready. The stars shined, and we pointed out constellations until one of us was stumped. We swapped war stories and laughs. I was taken back to my Afghan days, the good times, for a while. What a job and the money wasn't bad, either.

I eventually moved to mid-Missouri between Kansas City and St. Louis. My sister lives near KC, and I had friends living in Missouri. Colorado was a pricey place to reside. Mid-Missouri is peaceful, quiet, and lowkey. Quality Southeastern Conference college sports play in my backyard, and city fun is two hours in either direction. There's enough around to keep things interesting and a small-town vibe to keep me grounded. Things aren't perfect, but I feel easy, healthy, and free. My decisions and finances are stronger. I haven't ruled out a return to Georgia, but time will tell. But I'm sure that I'll move somewhere else at some point.

I'm entering my mid-forties, and most the guys I served with are in their early thirties. While reconnecting with my former platoon mates, I've learned that many have suffered similar issues to mine. They've had trouble fitting into the civilian world. Why? It's hard to put a finger on it. Statistics tell us that service with Uncle Sam places us at a greater risk for alcoholism, drug abuse, depression, homelessness, unemployability, and suicide. The government must do its best to care for warriors and avoid unnecessary conflict when possible. Warriors pay the price for the rest of their lives. Politicians don't. They thank us for our service yet understand nothing about what it entails.

Suicide has reached epidemic proportions among vets from the Iraq/Afghanistan era. We will know more veterans who took their own lives than died on the battlefield. I've had friends who've taken their own lives, and I wanted to go there myself. Studies tell us an average of twenty-two veterans end it each day, a national tragedy. There are several reasons why combat veterans pull the plug. I'm no expert, but I can speculate why through my own experiences and what I know of others.

"Had I died in combat, I would've gone out a hero. Had I died in combat, I wouldn't have become a drug addict and alcoholic. I wouldn't have gotten a DUI or caught a felony. I wouldn't be a loser."

In my opinion, those factors play a larger role in veterans' suicides than something like survivor's guilt. I've always said I'd trade places with Julian in a heartbeat. But the post-military legacy of heartache, pain, and failure led to my own suicidal thoughts more than the guilt of being a survivor. Perhaps if I followed my military years with immense success, I might feel as if I were honoring my fallen brother.

A person I served with, who attempted suicide, once sent me a text. It read. "I did well in high school. I never drank or did drugs, was an eagle scout, did varsity track and cross country, and achieved academic success. I joined the military and did exceptionally well. But, in late 2011, I had a pistol pointed at my head and was asking myself, 'Why shouldn't I pull the trigger?' because I'd fallen so far."

All in all, I endorse my dad's statement. "You couldn't give me a million dollars to do it again, but you also couldn't give me a million dollars not to have done it."

I got to serve with the greatest men I ever knew. As years have passed, contact with my old platoon mates is less frequent. Sometimes a text or phone call from a Third Platoon brother comes through, and we laugh, joke, and reminisce about the good ole days. Often, I'll receive an update about a brother I haven't spoken to in years. Sometimes the news is great, other times not so good, but I love hearing success stories.

We've often admired and emulated the men of the "greatest generation" because they fought fascism. Gen Xers and millennials will never be looked upon as the greatest. Slacker, entitled, spoiled, soft, and weak are words used to describe Generation X and Y. Thus, a quote I've heard several times applies to the veterans of the Global War on Terrorism. "We may not be the greatest generation, but we're the greatest of our generation."

We're the less than one percent who swore an oath and took up arms after the terrorist attacks on 9/11. My people are grunts. Most of us came from nowhere and had done little before volunteering for the infantry. We came from tough towns and tough backgrounds. A fair number of us were raised in middle-class America, so we didn't need a GI Bill to attend college. Instead, we loved our country and answered the call to defend it.

Perhaps the war in Afghanistan will go down in the annals of history as a failure. We're now learning that our leaders misled the public about accomplishments there and the prospects for future success. The fact is Afghanistan's government and military crumbled as soon as the US Military stopped helping. The powers-that- be said the government in Kabul would stand for months following a US Forces withdrawal. We grunts understood how incapable the Afghan National Army was and predicted they'd drop their weapons when we left their side. Once again, the suits didn't know what they were talking about. I'm not surprised everything ended in a clusterfuck. How's that for a twenty-year investment? Trillions spent, blood spilled, and lives altered for what?

General Douglas Lute mentioned in an interview that they (leadership) didn't know what they were doing in Afghanistan. That's the legacy of my war. Despite this failed legacy, I find solace in the fact that we selflessly fought for our country, our fellow man, and to better the lives of downtrodden Afghan people. As I write, no major attack has occurred on American soil since that horrible day on September 11. Osama bin Laden is dead.

Although these men are the best of their generations, the public will see Iraq and Afghanistan veterans at their worst. Thank God, I know how hard these men fought, how much they sacrificed, and the courage they displayed.

A decade after our war, my former platoon members are discovering their places in the world. Joey C and Harlen are roomies in Southern California, working and going to school. Joey C is still his happy-go-lucky self, albeit bearded and heavier, and he aspires to make movies. Currently, he manages a pot shop. Harlen works at the pot shop as a delivery driver. He's nearing graduation and becoming a possible leftwing activist, pitching democratic socialist ideas. Rusty, Dottie, and Hix also entered the world of academia. After marrying into an instant family with kids, Rusty got it together and earned his Bachelor of Science in Electrical Engineering from the University of Texas. Dottie completed his Bachelor of Science from West Virginia University and entered grad school. Impressive, especially for a young man who endured homelessness for a substantial amount of time. Hix obtained his Bachelor of Science in Environment and Society from the University of Alaska at Anchorage. He loves working in the great Alaskan wilderness. I was flattered when he told me that I'd inspired him to pursue a college degree.

Other men have sexy postwar stories. Smeed parlayed his infantry experience into a gig with the State Department as a civilian contractor and worked several additional tours in Iraq and Afghanistan. Ross worked private security details in the Middle East alongside Peshmerga fighters and served tours inside Africa. While at home in Colorado, Ross attends classes, works as a bouncer, and enjoys time as a newlywed. He aspires to open a brewery one day. Ole Stu is still happily married and loves spending time outdoors with his five kids. He makes a living doing risky work climbing towers. He also has a couple of side hustles, tattooing and crafting custom handgun holsters.

Some guys remained in the army and completed additional combat tours. T-Love, Mac, Marty, Harlen, and Phil deployed to Afghanistan again in December 2011. Phil completed ten years and three tours to Afghanistan before leaving the army and moving on to a career with the pipefitters' union. I'd bet civilian life is easier on his wife and four kids. Frankie served three combat tours and currently works as a recruiter near New York City. Shane continues to serve and, in an ironic twist, was sworn in as Charlie Company's first sergeant in the spring of 2020. I can't imagine how he felt when he parked in the 1SG spot for the first time and walked into his new office. He's an OG who'd grown up in Charlie Company. I have no idea where ole Rubio is. He lives off the grid. I'd bet that he's somewhere overseas gripping a rifle with a huge smile on his face. What I do know is after our return to Alaska, Rubio managed to get reassigned to another unit that was ramping up for a deployment to Afghanistan. The old warhorse is still galloping.

While living in Colorado, I hooked up with Gus and The Duke since they both lived near Denver. The Duke has a federal gig, a house in the burbs, and a wife and two kids. Gus has a lucrative sales career and is also happily married with two children. Nowadays, you can find Gus chilling by a pool with his fam; he moved to the Arizona desert. One night in Denver, Gus and I ran into Blackie as he was hoboing across the country. Blackie didn't paint a pretty picture of his life, and I didn't poke around being nosey. He's off-grid. I even crossed paths with my old infantry school bud, Willie B, since he too lived in Colorado. He earned a bachelor's from Colorado State and a master's from the University of Denver following his time in the army. He's now married and resides in Virginia.

My beloved Uncle Hook passed on while I was living the mountain life. He died on April 20, 2016 (420 to stoners). There couldn't have been a more appropriate time for the old hippie to cross over to the afterlife.

"Ole Iron Guts" Stan worked as an accountant in West Virginia and then moved to Oklahoma during an oil and gas boom. Crunching numbers never excited him; a patrol through the bush did. Following a stressful day, Stan will enter his garage and peek at the cheap Casio watch that he'd slid off the wrist of the dead Haji that he and Frankie smoked. It still beeps on the hour every hour.

On a sad note, in 2018, Brandon passed away at thirty-two. He was a squad mate, roomie, brother, and friend; his bravery was unmatched. A vision of him standing atop that MRAP, under no cover, firing away at the enemy, is engrained in my brain. Brandon was fearless, and he always had your back. The news of his death hit hard. But know this, Brandon lived it up during his short life. During most weekends in Alaska, he hunted ducks, fished for salmon, and took part in whatever outdoors activity was in season. He wasted none of his years. The only thing I heard about his passing was that he died peacefully in his sleep.

In 2021, I heard that ole Wimes had died at his own hand. This one hit hard, too. He died at the age of thirty-two – the age that I was when I met him in 2008. Words can't describe how much I'll miss his dancing around screaming, "It's Wime Time, baby!" These premature deaths must stop.

What about Doc Chauncy, our hero medic? He was released from prison after serving a decade. His debt to society has been paid, and he's living his best life.

I've found peace in my new life. It's been over ten years since I'd consumed anything harder than marijuana; I'm just a boring stiff living an average life. I'm on the way to being the old version of me, a social butterfly and happy-go-lucky. I'm back to being that laidback guy. The VA is now doing better by me, my finances are in order, and I'm living a better life. I won't lie; I still battle demons. But there are fewer demons to slay, and I'm more equipped to handle them.

While many folks my age are going through a midlife crisis, I'm experiencing a midlife blessing. Through all the ups and downs, I came out scarred but stronger, aged yet wiser. I've got no beef with the Taliban man, and I'm no longer tormented. The ladies have taken notice again, too. Ole Ronny B is back, baby, and I keep on *Rollin' with the Flow* like Charlie Rich. The old grunt has left his rifle in the arms room \m/.

GLOSSARY OF ACRONYMS AND TERMS

ABP – Afghan Border Police

ANA – Afghan National Army

AO – Area of Operations

ASG – Afghan Security Guard

ATLiens – The people of Atlanta, Georgia

AWOL – Absent Without Leave

BDA – Battle Damage Assessment

Big Chicken Dinner – Bad Conduct Discharge

Blue Falcon – Buddy Fucker

Boot – Marine fresh out of basic training

Cherry – Soldier fresh out of basic training

CIB – Combat Infantry Badge

COP – Combat Outpost

CROWS – Common Remotely Operated Weapon Station

DFAC – Dining Facility

DICK – Dedicated Infantry Combat Killer

DUSTWUN – Duty Status–Whereabouts Unknown

EOD – Explosive Ordinance Disposal

ETS – Expiration Term of Service

FO – Forward Observer

FOB – Forward Operating Base

Fobbit – Soldier who rarely, if ever, leaves the relative safety of the FOB

Gist – Intelligence heard by an interpreter scanning enemy radio traffic

GOAT – Greatest of All Time

HLZ – Helicopter Landing Zone

IED – Improvised Explosive Devise

IG – Inspector General

Joes – Junior enlisted soldiers

Kalat – Afghan home constructed with mud and rocks

KIA – Killed in Action

Klick—Kilometer

Latrine—Bathroom

LEG – Low Energy Ground

MEB – Medical Evaluation Board

MP – Military Police

MRAP – Mine-Resistant Ambush Protected vehicle

MRE – Meal Ready to Eat

Mustang – An officer who once served in enlisted ranks

MWR – Moral Welfare and Recreation

NCO – Non-Commissioned Officer

NVGs – Night Vision Goggles

OCS – Officer Candidate School

POG – Person Other than Grunt

POI – Point of Impact

POO – Point of Origin

POW – Prisoner of War

PT – Physical Training

PTSD – Post Traumatic Stress Disorder

PX – Post Exchange

QRF – Quick Reaction Force

R&R – Rest and Relaxation

Rear-D – Rear Detachment

REMF – Rear Echelon Motherfucker

ROE – Rules of Engagement

Romans – The people of Rome, Georgia

RPG – Rocket-Propelled Grenade

RTB – Return to Base

RTO – Radio Telephone Operator

Shura—Meeting where decisions are made in Islamic societies

SITREP – Situation Report

SOG – Sergeant of the Guard

Stop-Loss – The involuntary extension of a service member's active-duty status to retain them beyond their ETS

TC – Truck Commander

TOC – Tactical Operations Center

VA – Veteran's Administration

VBIED – Vehicle-Borne Improvised Explosive Devise

Wadi—Dried riverbed serving as road

Willy Pete – White Phosphorus

WTC – Warrior Transition Course

WTU – Warrior Transition Unit

CHARACTERS

Big Ronny B—Narrator's father of the same name

Blackie—Third Platoon, grenadier in narrator's squad (Second Squad)

Brandon—Third Platoon, SAW gunner in narrator's squad (Second Squad)

Captain Mac—Charlie Company Commander

Clover—"Old buddy" who smoked Afghan Kush with the narrator

Doc Lindley—Third Platoon, platoon medic during last half of deployment

Doc Rodrigo—Third Platoon, platoon medic during first half of deployment

Dottie—Third Platoon, machine gunner and bi-racial "rock and roller" (Weapons Squad)

Gus—Third Platoon, SAW gunner (Third Squad)

Frankie—Third Platoon, team leader, narrator's workout buddy at Malekshay (Third Squad)

Harlen—Third Platoon, eighteen and youngest in platoon (First Squad)

Hix—Third Platoon, acting platoon sergeant while Manny Fresh was gone (Weapons Squad Leader)

Joey C—met the narrator during training and joined him in Third Platoon

Julian—Third Platoon, grenadier, and unofficial chef at Malekshay (First Squad)

Manny Fresh—Venezuelan Third Platoon Sergeant

Paris—Second and Third Platoons, machine gunner from Memphis (Weapons Squad)

Parthia—Third Platoon, early Bravo Team Leader (Second Squad)

Phil—Third Platoon, "smack-talking" machine gunner with several kids (Weapons Squad)

Ross—Third Platoon, The narrator's meathead workout buddy (Third Squad)

Rubio—Third Platoon, gung-ho Puerto Rican staff sergeant (Third Squad Leader)

Sammy—Third Platoon, narrator's original platoon leader

Sergeant Chauncy—Head Charlie Company medic who was popular among the men

Shane—Third Platoon, messy Pennsylvanian workout buddy and team leader (First Squad)

Smitty—Third Platoon, SAW gunner in narrator's fireteam (Second Squad)

Stan—"Ole Iron Guts," Third Platoon, narrator's squad leader (Second Squad Leader)

Stu—Third Platoon, narrator's Alpha team leader (Second Squad)

The Duke—Third Platoon, team leader (First Squad)

Troutman—Third Platoon, narrator's platoon leader after Sammy

Wildfire—narrator's businessman friend from his hometown

Wimes—Third Platoon, high-strung SAW gunner (First Squad)

CPSIA information can be obtained
at www.ICGtesting.com
Printed in the USA
BVHW081554110523
663999BV00015B/981